DEWALT®

BUILDING CONTRACTOR'S LICENSING

Exam Guide

SECOND EDITION

By American Contractors Exam Services

Published by:

www.DEWALT.com/guides

Pal Publications, Inc.
800 Heritage Drive
Suite 810
Pottstown, PA 19464-3810
800-246-2175

Second edition published 2008

ISBN-10: 0-9797403-8-X
ISBN-13: 978-0-9797403-8-1
12 11 10 09 08 5 4 3 2 1
Printed in Canada

Titles Available From DEWALT:

Trade Reference

Blueprint Reading
Construction
Construction Estimating
Construction Safety/OSHA
Datacom
Electric Motor
Electrical Estimating
Electrical Professional Reference
HVAC Estimating
HVAC/R—Master Edition
Lighting & Maintenance
Plumbing
Plumbing Estimating
Residential Remodeling & Repair
Security, Sound & Video
Spanish/English Construction Dictionary—Illustrated Edition
Wiring Diagrams

Exam and Certification

Building Contractor's Licensing
Electrical Licensing
HVAC Technician Certification
Plumbing Licensing

Code Reference

Building
Electrical
HVAC/R
Plumbing

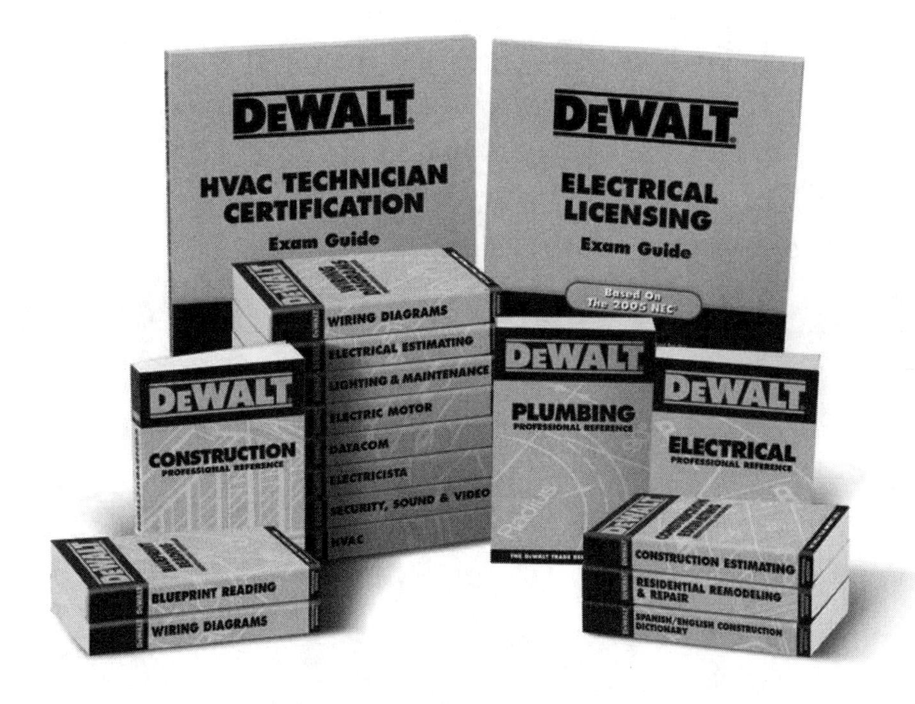

www.DEWALT.com/guides

CONTENTS

DEWALT BUILDING CONTRACTOR'S LICENSING EXAM GUIDE – SECOND EDITION

by
Chris Prince
American Contractors Exam Services

INTRODUCTION

Congratulations! If you are reading this book, it is likely that you are pursuing new opportunity. Whether you are preparing to take an exam in order to start your own contracting business, to enhance the qualifications of an existing company, or to qualify the company for which you now work, you are taking deliberate measures to improve an existing situation. So sit back, relax, and rest assured that the time you spend with this book will significantly improve your chances of passing the exam that you are about to take.

TIME FOR TEST MODE!

Preparing for and taking an exam requires a specific mind-set. It requires planning for the upcoming event and making a conscious effort to control anxiety. For most, the thought of taking an exam will cause a feeling of butterflies churning in the stomach. For some, it will lead to more severe symptoms such as headaches and nausea. Don't fear; controlled test anxiety is good. It creates a sense of urgency to succeed. Identify your level of anxiety, and control it.

How do you rate your level of anxiety? For some, perhaps the last test-taking ordeal was a high school final exam twenty years or so ago. The memory of this experience, coupled with the pressure of living up to the expectations of family, a boss, or coworkers can be overwhelming. When children are aware that a parent will soon be taking an exam, they are likely to reciprocate the pressure applied to do well in school. Bosses often assume that the exam will be a breeze for such a competent employee, and many times, a spouse's support is mistaken for unrealistic expectation. Peer pressure is not necessary. Do what you can to minimize it.

Keep the fact that you are taking the exam a secret! Maybe everyone notices your study efforts, but there is no need to announce the test date. Make your plans by scheduling the exam, and if you must explain, say that you have an appointment. The pressures felt from the need to please and live up to the expectations of everyone around you can lead to undue stress and is not necessary. If no one knows your plans, the only pressure you will feel to succeed is your own. This pressure is healthy and probably necessary. Rather than dealing with the pressure of others, imagine the feeling when you announce your accomplishment after the fact. Work toward this goal, and visualize success. Think positive.

Another way to eliminate stress may be to think of your first test date as a practice run. In the worst-case scenario, passing the exam will take two attempts. It is not the end of the world if you do not pass on the first attempt. Most states and municipalities allow you to take the exam as many times as necessary without penalty. Before using this suggestion, make sure that your exam does not limit the number of times it can be taken during a certain amount of time. (We will explain later how to check this.)

Notes

Next, avoid becoming irritated that the exam is required. Forget the fact that you know your trade and feel that you should not be required to pass an exam to prove yourself. Look at the experience as an opportunity. The knowledge gained from preparing for the exam will far outweigh the effort. Without a doubt, you will gain information that will improve your efficiency and trade expertise, and could even save you money.

Remember the old carpenter's advice to measure twice and cut once? Similarly, when it comes to test-taking, it is critical to be precise, careful, and methodical as you prepare for exam day. Apply this philosophy as you take the practice exams in this book—read the questions twice and choose the answers carefully. The exams are typically administered in a multiple-choice format, and it is not uncommon for at least two of the choices to appear to be correct, depending on how the question is read. Remember, there is only one correct answer, and one word in the question can change its entire meaning. Be careful.

After you have switched to test mode, it is time to prepare for the exam.

PREPARING FOR THE EXAM

The first step of preparation is to identify the exam required to accomplish your goal. If you are taking a state-required exam to become a licensed contractor, you can begin by visiting www.becomealicensedcontractor.com. Select your state, choose General Contracting under the listing of test types, and the test requirements will be found. The site will likely lead you to the candidate information bulletin provided by the approved testing company.

The candidate information bulletin is the test-taker's rule book. It provides specific information, including the addresses of testing centers, identification requirements, security procedures, and anything else necessary for taking the exam. In preparing for the exam, it is important to identify the answers to the following questions when using the candidate information bulletin.

1. ***What are the approved references?*** This is, by far, the most important piece of information to know prior to your exam preparation mission. Most general contracting exams are based on either the International Residential Code or the International Building Code and several additional references. After identifying the approved reference materials, eliminate the study exams in this book that are not included in your exam. Do not spend valuable time studying information that is not included on your exam.

Notes

2. ***Are the approved references allowed in the examination center?*** In most cases, books are allowed in the examination room for use throughout the exam. While this would seem to make the exam extremely easy, if you are not familiar with how to use the books, it might as well be a closed-book exam.

3. ***Is tabbing and highlighting of the books allowed?*** Typically, if the approved references are allowed to be taken into the exam room, they are allowed to be highlighted and tabbed prior to test day. Many rules allow you to underline but prohibit you from making notes in the reference materials. The candidate information bulletin will specifically call for permanent tabs. Permanent tabs are those that cannot be easily removed. Post-it tabs are generally not allowed. If any of the rules are broken, your reference materials may be banned from the test site.

4. ***What is the time frame for taking the exam?*** The time frame is important because you will want to simulate the time allowed as you work through the study exams throughout this book. Simply take the allotted time in minutes and divide by the number of questions on the exam to determine the time allotment per question. You can now easily multiply the number of questions in a particular section of this book by the average time allowed per question to set the time frame for working through the practice exams. In the beginning, you will want to allot additional time until you become familiar with the reference materials.

5. ***What is the content outline of the exam?*** It is important to pay attention to the breakdown of the exam in order to allot your study time. If only five out of 80 questions pertain to OSHA, you should spend less time studying this subject and concentrate more on the areas representing the largest portion of the exam.

USING THIS EXAM GUIDE

Before using this guide, make sure that you have highlighters and tabs readily available. Begin with identifying the table of contents, the index, and the glossary of each reference book. Place a tab on each of these sections for easy access. The table of contents divides the book into chapters or subjects and will be used frequently. The index is an alphabetical listing of key words found throughout the book and should be your starting point for finding an answer to a specific question. If the answer is not found using the index, identify the chapter according to the subject of the question by using the table of contents. The glossary is an alphabetized listing of terms that serves as a useful source for answering questions

Notes

quickly. A glossary may or may not be found in the approved references.

Each section of this exam guide is based on a particular subject. As you answer each of the questions from the study exam, highlight the answers. This will help you to become familiar with the book and will strengthen your ability to quickly reference important code items. Remember, the purpose of each practice exam is not to test your trade knowledge; it is to provide an exercise of how to navigate and use the approved reference(s). If you feel that you know the answer to a question, you should verify the answer using the applicable book. Pay close attention to the tables that are referenced as well as the subject matter of each chapter.

When it comes to tabbing your reference materials, be careful not to overdo it. Placing too many tabs will be more of a hindrance than a useful tool. Remember, when using tabs, the more tabs you use, the longer it will take to read through each one. Using the index will likely save you time in your search for specific information. It is highly recommended that you tab the chapters, table of contents, index, and useful tables.

Going into the exam, make sure that you are familiar with the subject matter of each book. It is imperative that you understand the layout and format of each of the references. This is accomplished through the process of answering the questions in this guide.

Prior to exam day, remove the index from approved reference books that are bound with a three-ring binder, staple it together, and place it in the front pocket of the binder. This will allow easy access to the index by being able to place it on the testing table beside the reference book rather than requiring you to flip back and forth between the index and the body of the book.

On average, exams allow about 3 minutes per question. When you begin using the study exams in this book, allow yourself 8 to 10 minutes per question. As you become more familiar with the reference materials, decrease this time allotment to about 5 minutes. By the time you complete the final exams, only allow yourself the time allotted on your exam.

It is important to devise a time management strategy that works for you immediately. By test day, your goal is to have a plan of action of how to work through the exam. As you are preparing for the exam, keep in mind that the clock will be working against you. Decide the maximum amount of time that you will spend on one question before moving on to the next. You do not want to run out of time.

Notes

WHAT TO BRING TO THE EXAM

Prior to test day, make sure you get a good night's rest and arrive armed with the following items:

Candidate information bulletin. Many times, the proctor of the exam is inexperienced. If you are told that tabs are not allowed, you need to be able to defend yourself by referring to the bulletin (the rule book).

Bottle of water. The clock does not stop during the exam. There are no "hold" buttons. If you need a sip of water and have to run to the water fountain or bathroom, the timer will continue to count down. Although you have little choice when it comes to a bathroom break, at least be prepared for the dry mouth syndrome.

Magnifying glass. Many of the documents and diagrams used throughout the exam are difficult to read. Save the frustration and headache-causing eyestrain, and use a magnifying glass.

Two pencils and a pencil sharpener. Arrive with at least two pencils and a pencil sharpener, especially if you are taking the exam using the old-fashioned pencil-and-paper format. Tests are still administered this way in several states and municipalities.

Two calculators. Remember Murphy's law—if it can go wrong it probably will on test day. Have a contingency plan for everything. If you insist on using your favorite calculator, and it happens to be one that you are not sure is allowed, such as a construction master, make sure that you have a backup.

A great attitude. Make every attempt to remain calm, cool, and collected. This is easier to maintain if you have had adequate rest the night before the exam. Do not cram and stay up until midnight. This will work against your tolerance level for aggravation. Remember, frustration will only create tension and make everything more difficult.

Make sure that you also understand the requirements for identification, payment methods, and proper exam registration documentation.

LET THE TEST BEGIN

The moment you sit in the "hot seat" to begin the exam, let the strategy unravel. Have a plan and stick with it. A few recommendations:

1. ***Switch to test mode.*** Many times, the rules of thumb and assumptions you make in the field will not work in the test world. While your background and experience can be an attribute, do

Notes

not allow it to get in the way on test day. Make no assumptions. If you are not 100% sure of an answer, try to verify it at some point. Be deliberate to focus and concentrate on each question.

2. ***Arrange your work area neatly.*** Stack your books to one side and, if possible, stand them upright so that they are easily accessible. Decide for yourself that at least for this day, you will be the most organized and careful person in the world.
3. ***Download the memory.*** That's right, download the information you are having trouble remembering. Transfer the formulas and anything else you have been repeating since you walked into the test site from your brain to the scratch sheet of paper provided by the proctor.
4. ***Become familiar with the construction drawings and/or diagram booklet before you start the clock.*** If your test is being administered by computer, you are in control of when the countdown begins. Take advantage of the control, but be careful not to push it. The proctor is only a few steps away and may prod you to begin the test if you wait too long.
5. ***Answer the easy questions first.*** Nothing will boost your confidence more than to run through a good portion of the exam answering questions based on information you recall from your studies. In contrast, your confidence level can diminish rapidly if you become distracted by a difficult question.
6. ***"Mark" any questions answered that you doubt.*** If the test is computer based, you have the option to mark questions to review later. If you run out of time, the computer accepts the selected answer and does not penalize you for marking the question. By marking questions, if you have additional time after answering all questions, you can verify the answers you selected.
7. ***Leave the difficult questions unanswered, and come back to these last.*** As you skip questions, make a note on the scratch paper, associating the question number with a specific book. This will allow you to categorize each of the unanswered questions by book, saving valuable time and unnecessary frustration. After reading the last question, you will be allowed to revisit unanswered questions or marked questions. You can choose to go to the first of these in the group or to a specific question number.
8. ***Do not run out of time.*** Pay attention to the clock. Do not leave any question unanswered. Before you run out of time, select a

Notes

choice for each question. Questions left unanswered will be counted against you. If you have to guess on a number of questions, improve your odds by selecting the same choice on each question. For example, select all "a's" or all "c's," but your goal is to manage your time and not have to guess on any of the questions.

STATE LICENSING REQUIREMENTS FOR GENERAL CONTRACTORS					
State	Licensing Board Phone Number Company	Licensing Board Website	State Licensing Exam/Testing	Pre-licensing or Pre-approval	Continuing Education
Alabama (Commercial only)	(334) 272-5030	www.genconbd.state.al.us	Yes	Yes/Pre-approval	No
(Residential only)	(334) 242-2230	www.hblb.state.al.us	Yes	No	No
Alaska (General Contractor)	(907) 465-3035	www.dced.state.ak.us/occ/pcon.htm	No	No	No
(Residential only)	(907) 465-3035	www.dced.state.ak.us/occ/pcon.htm	Yes/ Thomson Prometric	No	Yes
Arizona	(602) 542-1525	www.rc.state.az.us/	Yes/Thomson Prometric	No	No
Arkansas	(501) 372-4661	www.state.ar.us/clb	Yes/ Thompson Prometric	No	No
California	(800) 321-2752	www.cslb.ca.gov	Yes/ California State Licensing Board	Yes/Pre-approval	No
Colorado	Not Regulated by State/contact city/county		No		
Connecticut	(860) 713-6135	www.dcpaccess.state.ct.us/	No	No	No
Delaware	(302) 577-8656	www.state.de.us	No	No	No
Florida	(850) 487-1395	http://www.myflorida.com/dbpr/	Yes/ Professional Testing Inc.	No	Yes
Georgia	(478) 207-1416	www.sos.state.ga.us/plb/construct/	Yes/PSI	Yes/Pre-approval	No
Hawaii	(808) 586-2689	www.hawaii.gov/dcca/pvl	Yes/Thomson Prometric	Yes/Pre-approval	No
Idaho	(208) 334-3233	https://www.ibol.idaho.gov	No	No	No
Illinois	Not Regulated by State/contact city/county		No		
Indiana	Not Regulated by State/contact city/county		No		
Iowa	(515) 281-7995	www.iowaworkforce.org/	No/Register with State; City/County may require testing		
Kansas	(785) 368-8222	www.ksrevenue.com	No/Register with State; City/County may require testing		
Kentucky	Contact City/ County		No/Register with State; City/County may require testing		
Louisiana	(225) 765-2301	www.lslbc.state.la.us	Yes/Louisiana State Licensing Board	Yes/Pre-approval	No

(continued)

STATE LICENSING REQUIREMENTS FOR GENERAL CONTRACTORS (*Continued*)					
State	**Licensing Board Phone Number Company**	**Licensing Board Website**	**State Licensing Exam/Testing**	**Pre-licensing or Pre-approval**	**Continuing Education**
Maine	Not Regulated by State/contact city/county		No		
Maryland (Commercial only)	Register with city/county		No		
(Residential only)	Register with State Attorney General's office		No		
Home Improvement	(410) 230-6309	www.dllr.state.md.us	Yes/PSI	No	No
Massachusetts	(617) 727-7532	www.state.ma.us/bbrs/hic.htm	Yes/ Thomson Prometric	No	
Michigan	(517) 241-9254	www.michigan.gov/dleg	Yes/PSI	No	No
Minnesota (Residential)	(651) 284-5065	www.doli.state.mn.us	Yes/Promissor	No	No
Mississippi	(601) 354-6161	www.msboc.state.ms.us	Yes/ PSI	Yes/Pre-approval	No
Missouri	Not Regulated by State/contact city/county		No		
Montana	(406) 444-7734	www.mtcontractor.com	No/Register with State		
Nebraska	(402) 595-3095	www.dol.state.ne.us	No/Register with State		
Nevada	(702) 486-1100	www.nscb.state.nv.us	Yes/PSI	Yes/Pre-approval	No
New Hampshire	Not Regulated by State/contact city/county		No		
New Jersey	(609) 984-7910	www.state.nj.us	No/Register with State		
New Mexico	(505) 452-8311	http://www.rld.state.nm.us/CID/index.htm	Yes/PSI	Yes/Pre-approval	No
New York	Not Regulated by State/contact city/county		No		
North Carolina	(919) 571-4183	www.nclbgc.org	Yes/PSI	Yes/Pre-approval	No
North Dakota	(800) 352-0867 ext. 83665	www.state.nd.us/sec	No/Register with State		
Ohio	Contact Local Building Department		No		
Oklahoma	Not Regulated by State/contact city/county		No		
Oregon	(503) 378-4621	www.ccb.state.or.us	Yes/PSI	Yes/Pre-licensing	No

Pennsylvania	Not Regulated by State/contact city/county		No/city/county may require testing		
Rhode Island	(401) 222-1268	www.crb.state.ri.us	No/Register with State		
South Carolina (General Contractor)	(803) 896-4686	www.llr.state.sc.us	Yes/PSI	No	No
(Residential only)	(803) 896-4696	www.llr.state.sc.us	Yes/PSI	Yes/Pre-approval	No
South Dakota	Not Regulated by State/contact city/county		No		
Tennessee	(615) 741-8307	www.state.tn.us/commerce/boards/contractors/	Yes/PSI	No	No
Texas	Not Regulated by State/contact city/county		No		
Utah	(801) 530-6628	www.dopl.utah.gov	Yes/Thomson Prometric	Yes/Pre-approval	No
Virginia	(804) 367-8511	www.state.va.us/dpor	Yes/PSI	Yes/Pre-licensing	No
Washington	(800) 647-0982	www.lni.wa.gov	No/Register with State		
West Virginia	(304) 558-7890	www.labor.state.wv.us/	Yes/Thomson Prometric	No	No
Wisconsin	(608) 261-8500	www.commerce.wi.gov	No/Register with State		
Wyoming	Not Regulated by State/contact city/county		No		

For more information visit: www.examprep.org

Notes

PART ONE
Math Concepts

Understanding basic math concepts can be crucial for the portion of the exam, which is based on print reading.

If your candidate information bulletin does not list math or plan reading as subjects to be covered, this chapter may not be as important to you as some of the others. However, it will prove beneficial for shifting your thought process to test mode. Working through the exercises will stimulate your brain. This chapter can be thought of as the "warm up" recommended for most athletes prior to an intense workout.

While many contractors probably dread math as much as many Americans dread exercise, it is as crucial to running a successful business as exercise is to health. What many exam candidates come to realize is that math is not nearly as bad as what they have come to expect. In fact, most people are better at math than they think.

The following chapter covers basic math concepts that are crucial to the contracting business and, more importantly, for now at least, crucial to passing an exam. While it is impossible to cover every type of question that may appear on an exam, a strong knowledge of the proper steps to performing similar math calculations will get you through any exam.

Concepts covered in this chapter include:

- Math rules
- Rounding
- Converting inches to feet
- Square feet to estimate brick, block, and drywall
- Cubic yards to estimate concrete and dirt
- Framing math
- Roofing math

Notes

THE DREADED MATH

When it comes to math portions of the exam, there are several important and critical rules. It is in your best interest to convince yourself that for this one day, you are going to take a methodical approach and deliberately follow each of the rules that follow.

MATH RULES

1. ***Always use a calculator.*** This is probably the most critical rule when it comes to performing math operations on the exam. Never perform mental calculations. Once you arrive at an answer, do the calculation again to check your answer. Repeat your calculations until you are certain the answer is correct.
2. ***Use scratch paper to write down each step of the math.*** Professionals who use math on a daily basis, accountants, engineers, and the like, are sometimes the most difficult math students. Why? Because they simply refuse to use scratch paper to write each step of the equation as they work through it. They are also reluctant to use a calculator for what they believe to be simple math. You do not want to fail this exam by one or two points; it hurts worse to make a grade of 69 than it does to make a grade of 30. The point is, don't get in a hurry and miscalculate. Always write your problem out as you solve it. Always perform math operations, at least twice, with a calculator.
3. ***Always multiply "like" numbers.*** In other words, do not mix feet and inches. If you multiply 10 feet by 8 feet 6 inches, the answer is not 86 feet, it is 85 feet. You cannot simply multiply 10 by 8.6—you must first convert the 6 inches to feet (to be reviewed later).
4. ***When converting inches to feet, round to the nearest hundredth.*** Yes, this is an exception to Rule 3. The reason we round to the nearest hundredth when converting inches to feet is because it puts your answer in the correct range of the choices provided and allows you to write it down more easily. You would not want to write 0.3333333333333 when it would be much simpler to write 0.33 (to be reviewed later).
5. ***Never round until you get to the end.*** This simply means that you should never round any of your numbers until you solve the equation. When necessary, round based on the choices provided to a question or to the directions provided in the question. The only ***exception*** to this rule is when converting inches to feet (Rule #4).

Example:

Rounding up to the next whole yard, how many cubic yards of concrete will be needed to pour a 4-inch-deep drive that measures 20 feet × 100 feet?

a. 24
b. 24.44
c. 25
d. 25.5

Solution:

$$CY = \frac{L \times W \times D}{27}$$

$$CY = \frac{100 \times 20 \times .33}{27}$$

$$CY = 24.44$$

Explanation:

The question clearly states "round to the next whole yard." Although the exact answer is one of the choices, 25 (choice c) is the correct answer. Notice that 4 inches was converted to feet allowing the multiplication of "like" numbers. Before solving the equation, 0.33 was rounded. This is acceptable and will put you in the proper range of the correct answer for your exam.

6. ***Draw it out.*** If you have to sketch apples and stick people, do it. Remember, you do not want to fail this test by one question. It is critical that you use your scratch paper to draw as you solve the equation to answer the question.

MATH REVIEW

Rounding

Let's look at the concept of rounding. The numbers to the right of the digit, or point, begin with the tenth place. The second number is the hundredth place, and the third number is the thousandth place.

To round a number, begin by identifying the rounding digit, and look to its right side. If the digit to the right of your rounding digit is 4 or less, do not change it. All digits (numbers) to the right side of the rounding digit will be dropped.

Example: 0.3333 rounded to the nearest hundredth will be 0.33

What if the number to the right of your rounding digit is not 4 or less? If the digit is 5, 6, 7, 8, or 9, your rounding digit is increased by one number. All digits to the right side of the rounding digit will then be dropped.

Example: 0.6777 rounded to the nearest hundredth will be 0.68

Notes

Rounding to the nearest 100th

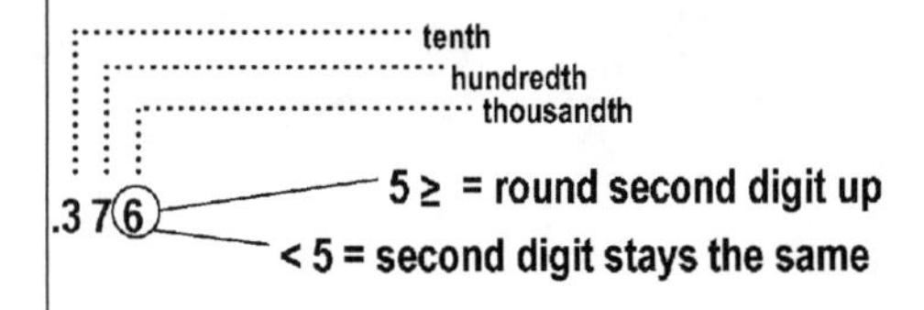

Notes

For exam purposes, rounding is used most commonly when converting inches to feet (next lesson) for the purpose of multiplying and/or dividing like numbers. Since each of these steps will be written on a sheet of scratch paper, rounding to the nearest hundredth will save valuable time and will lead you to choose the correct answer.

If you do not round to the nearest hundredth, you will still arrive at the correct answer, but you are more likely to make a mistake.

Converting Inches to Feet

Converting inches to feet is necessary for multiplying like numbers and will frequently be the first step required to solve a math-related question. Begin to understand this concept by stating to yourself "36 inches is equal to 3 feet." Think about it. If someone asks how many feet are in 36 inches, your reply will be almost instant. This may be handy on test day.

Begin to think of inches in terms of fractions. To do this, 36 inches would be placed over 12 inches ($^{36}/_{12}$).

To convert a fraction to a decimal, you simply take the top number (numerator) and divide by the bottom number (denominator). Thirty-six inches divided by 12 equals 3. Apply this to any number presented as inches, and you can quickly convert it to feet.

Four inches converted to feet would simply be 4 divided by 12 ($^{4}/_{12}$). This equals 0.3333333333. Rounding to the nearest hundredth, 4 inches is equal to 0.33 feet.

Practice:

1. Rounding to the nearest hundredth, 14 inches is equal to ____ feet.

   a. 1.4

   b. 1.2

   c. 1.17

   d. 1.166

2. Rounding to the nearest hundredth, 8 inches is equal to ____ feet.

   a. 0.8

   b. 0.67

   c. 0.6666

   d. $^{8}/_{10}$

3. Rounding to the nearest hundredth, 3 inches is equal to ____ feet.

   a. 0.25

   b. 0.33

   c. 0.4

   d. 0.5

Answers/Solutions found in Part 17 of this book.

Applying the Concepts of Rounding and Converting Inches to Feet

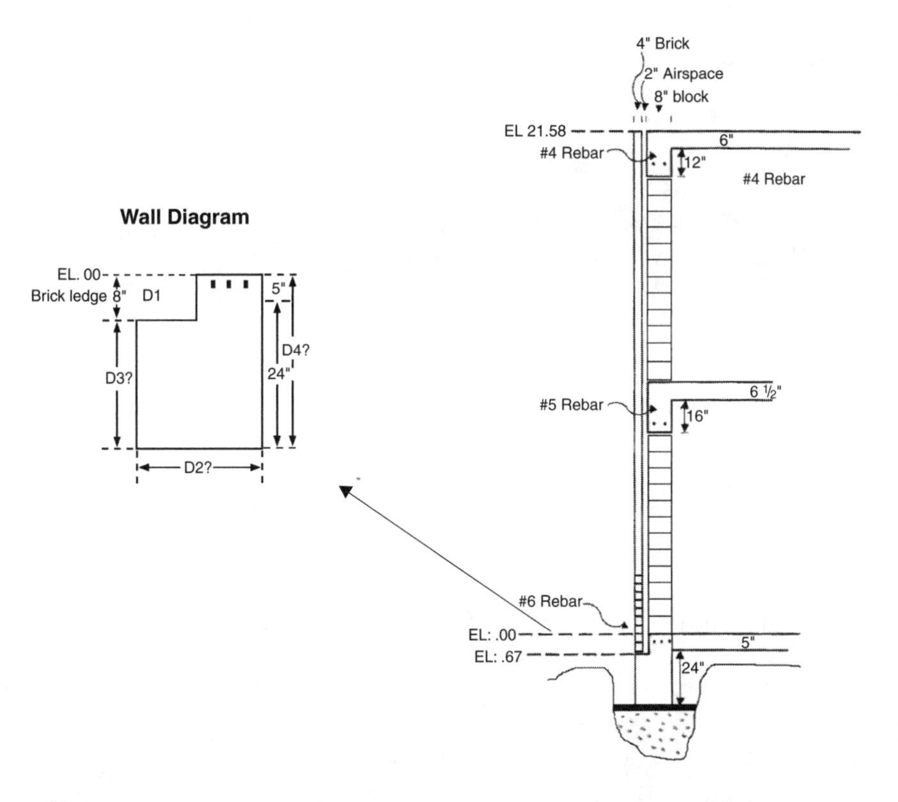

The drawing shown represents the cross section of a two-story building. A *cross section* shows a view of the internal construction of an object or structure. It represents how it looks when cut on a plane.

The footing, slab, brick ledge, and building height can be determined from the drawing. A *footing* is a base or bottom of a foundation, pier, wall, or column that can be manually formed (stem wall) and supported by an earth-formed footing.

This particular footing will support a 4-inch brick and an 8-inch block separated by 2 inches of air. All heights for this building will be relative to the slab level and will be represented by an elevation number.

Notes

Notes

An *elevation* is the height above an established reference point, such as the slab level in the drawing. The elevation at the top of the slab is identified as 0. The level portion of the footing that will support the brick is identified as –67. This indicates the brick ledge is 67 hundredths of a foot lower than the floor level.

Answer the following questions based on the Wall Diagram on page 13.

1. Calculate the depth of the brick ledge. (D1(?))

   a. 0.6 ft.

   b. 0.67 ft.

   c. 0.8 ft.

   d. 0.85 ft.

2. What is the width of the footing? (D2(?))

   a. 1.2 ft.

   b. 1.17 ft.

   c. 1.4 ft.

   d. 1.25 ft.

3. What is the outside height of the footing? (D3(?))

   a. 2.1 ft.

   b. 1.2 ft.

   c. 1.75 ft.

   d. 1.9 ft.

4. What is the inside dimension, from grade to the top of the slab, for the footing? (D4(?))

   a. 2.9 ft.

   b. 2.42 ft.

   c. 2.35 ft.

   d. 2.25 ft.

Calculate Square Feet

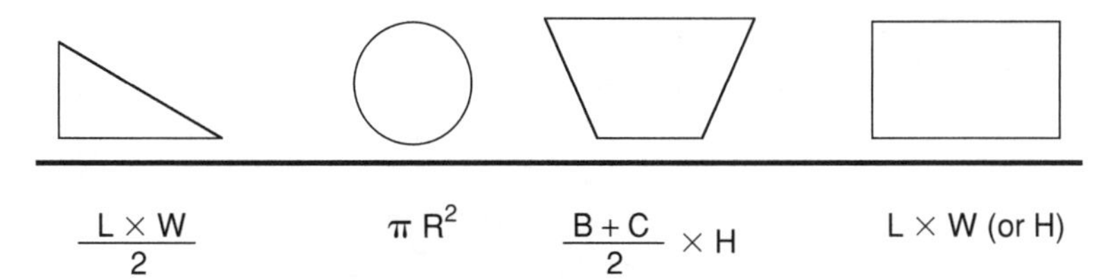

A majority of building components, including drywall, bundles of shingles, roof sheathing, brick, and block, are estimated by determining the square feet of the area first. Think of square feet as the foundation, or the starting point, for any take-off. If you understand how to properly calculate square feet, you can correctly answer many of the questions on the exam.

As discussed previously, you should be precise with your calculations. For example: If you are answering a question such as how many square feet of concrete slab will be installed, first determine the precise dimensions of the slab. Do not simply take the outside dimensions that have been provided. Read the details to determine where the slab is poured. Is it to the inside of the building, or does the building sit on top of the slab?

Example:

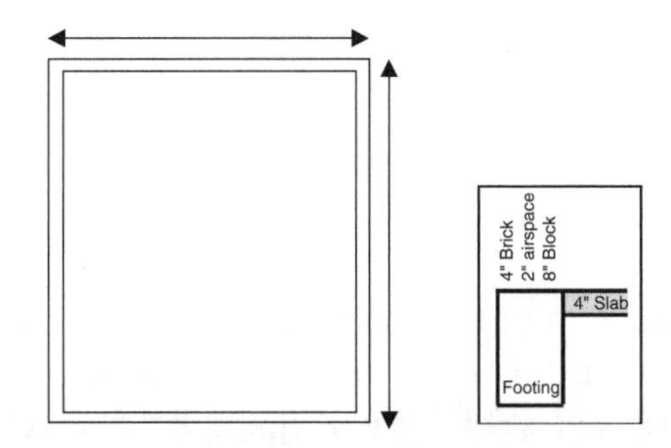

What is square footage of the 4-inch slab if the building dimension is 31 feet 6 inches × 20 feet?

a. 632 sq. ft.

b. 630 sq. ft.

c. 583 sq. ft.

d. 515 sq. ft.

Explanation:

The width of the footing must be subtracted from the overall dimensions of the building because the detail specifies that the slab is poured to the inside. The width is 14 inch, which converts to 1.17 feet.

20' − 1.17 − 1.17 = 17.66 (subtract the thickness twice)
31.5 − 1.17 − 1.17 = 29.16 (subtract the thickness twice)
17.66 × 29.16 = 514.9656 (answer is (d))

Notes

Notes

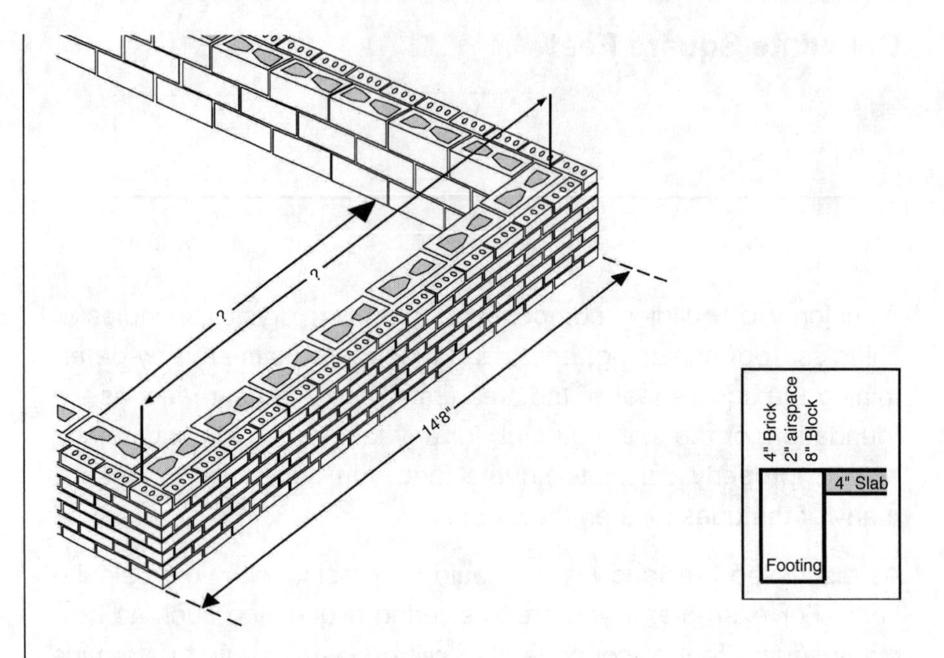

5. Based on the information provided, what is the dimension when measured on the inside of the structure?

   a. 14.67'

   b. 12.67'

   c. 13.5'

   d. 12.33'

6. If the construction of this building did not include brick and the dimensions of the interior remain the same as determined in Question #1, what is the measurement of the structure?

   a. 12.67'

   b. 13.67'

   c. 13.5'

   d. 11.67'

7. If this building were rectangular and measured 14 feet 6 inches in width and 23 feet in length, what is the inside dimension lengthwise?

    a. 22.67'

    b. 22'

    c. 21.83'

    d. 20.66'

Notes

Using Area (Square Feet) to Estimate Sheathing, Drywall, Brick and Block

To estimate quantities of flat material, simply divide the area to be covered by the square feet of the material to be installed. For example, if the square feet of the interior walls for a structure totals 3,200, and 4' × 12' × ½" gypsum is being installed, divide 3,200 by 48 to determine that 67 boards must be ordered.

An obvious question at this point might be "Do I deduct for openings?" Do not deduct for openings when calculating gypsum/drywall. Do, however, deduct for openings when calculating brick or block.

1. How many sheets of 4' × 8' × ⅝" gypsum board will be needed for the interior of a garage with room dimensions of 24' × 30' with 8' ceilings? *Note:* The gypsum is to be installed on the walls and ceiling. Disregard openings.

    a. 23

    b. 27

    c. 50

    d. 69

2. After adjustments have been made to account for the pitch, a roof is determined to be 2,875 square feet. How many 4 × 8 pieces of sheathing will be installed?

    a. 90

    b. 60

    c. 120

    d. 52

Notes

3. Half-inch gypsum board is to be installed on the 8' walls of a room with dimensions of 12' × 14'. Specifications call for 4' × 8' × ⅝" gypsum board to be installed on the ceiling. The room has an opening of 6' on one side with two windows, each 4' × 6' facing the front of the structure. How many 4 × 12 sheets of ½" gypsum must be ordered?

   a. 13

   b. 9

   c. 5

   d. 4

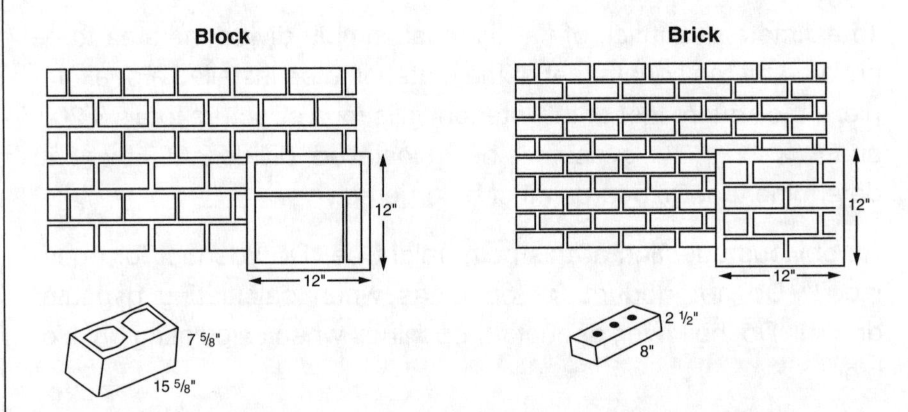

Estimating brick and block is similar to estimating gypsum or sheathing. Begin by determining the face area of the material to be installed. For example, if an 8" × 16" nominal block (7⅝" × 15⅝" plus the mortar joint) is to be used, the face area of the block is 128 square inches. To convert this to a multiplier, you can follow one of several methods.

First, you can divide 144 square inches by 128 square inches to determine that it will take 1.125 blocks per square foot. The area (square feet) multiplied by 1.125 determines the number of 8" × 16" block needed.

You can also divide 128 square inches by 144 square inches to determine that one block will cover 0.8888 square feet. The area (square feet) divided by 0.8888 will determine the number of 8" × 16" block needed. Either way will result in the same answer.

Many contractors take the length of a wall and divide by 1.33 to determine the number of block required on each course. They will then take the height of the wall and divide by 0.67 (height of an 8 inch block) to determine the number of courses required. By multiplying the number of courses by the number of blocks on each course, the total quantity needed for the structure can be determined.

For test purposes, it is recommended that you follow one of the first two methods. Begin by determining the total area in square footage and multiplying/dividing by the appropriate factor.

Example:

How many blocks are needed for a wall that is 21 feet tall and 38 feet long?

a. 898

b. 600

c. 917

d. 1,191

Explanation:

Determine the area of the structure by multiplying the height times the length: 21' × 38' = 798 square feet. The area divided by 0.8888 will determine quantity of block needed: 798 ÷ 0.8888 = 898 block. The area multiplied by 1.125 will also determine the quantity of block needed: 798 × 1.125 = 898 block. The answer is (a).

To estimate the quantity of brick needed for a structure, follow the same steps. However, we will look at only one way to determine the correct answer. Begin this task by determining the number of brick needed for each square foot of area. Next, determine the area, and multiply the area by the multiplier to arrive at the correct answer.

If a brick is 2½" × 8", the face area of the unit is 20 square inches. The square inches of a square foot (144) divided by 20 equals 7.2 brick. If the area to be constructed with a brick veneer is 1,000 square feet, multiply the area by 7.2 to determine that 7,200 brick should be ordered for this project.

At this point, you may be thinking of waste. Factors will be provided on your exam to determine the allotted waste. If the waste factor on this project (question) is specified as 15%, simply add this amount to the total: 15% of 7,200 is 1,080; 7,200 plus 1,080 is 8,280. If this were a question on the exam, the correct answer would be 8,280 brick.

4. Disregarding waste and loss due to overlapping at the corners, how many block would be needed for a 20' × 32' building that is 18' tall?

   a. 1,872

   b. 2,106

   c. 2,489

   d. 2,725

Notes

Notes

5. Based on a waste factor of 10%, how many brick will be needed for a 10' × 20' wall? The brick is 7" × 2¾".

   a. 1,295

   b. 1,485

   c. 1,646

   d. 1,760

6. A wall is 32' long, 8' high, and will be constructed out of 8" × 8" × 16" concrete masonry units (CMU) with a brick veneer. Total square feet of openings is 28, and the brick is 2½" × 8". How many of each is required?

   a. 1,642 brick, 257 block

   b. 1,487 brick, 228 block

   c. 1,287 brick, 188 block

   d. 1,189 brick, 171 block

Calculate square yards

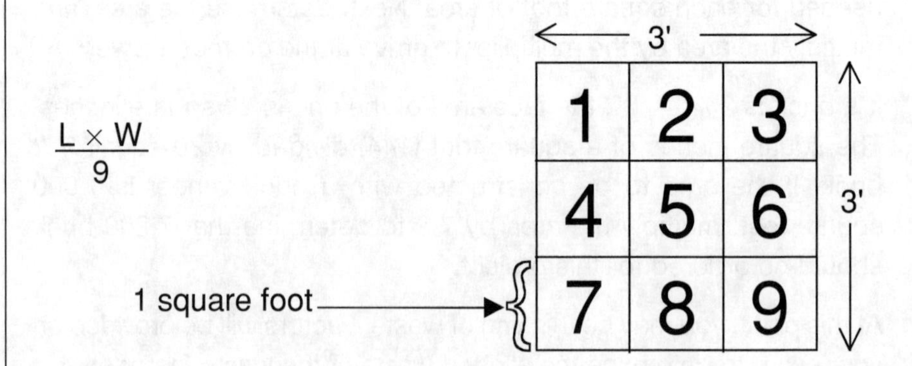

To calculate square yards, divide the area (square feet) by 9. There are 9 square feet in 1 square yard. It is highly recommended you always write the formula down before you begin to solve the equation. Write the formula down first, and carefully insert each number from the question into the formula.

It is not uncommon for an exam question to include too much information. In the testing profession, this information is referred to as a *distracter*. By paying attention to the formula, the distracter is less likely to cause a mistake in your calculations.

Example:

How many square yards of asphalt will be installed for a driveway that is 12 feet wide and 42 feet long? The depth of the asphalt is 2 inches.

a. 85.68 sq. yd.

b. 56 sq. yd.

c. 112 sq. yd.

d. 3.17 sq. yd.

Explanation:

Choices are provided for nearly any mistake made using the 2 inch depth. The proper way to calculate the square yards for this question is to calculate 12×42 and divide by 9. $12 \times 42 \div 9 = 56$. The answer is (b).

1. Calculate the square yards for a 20' × 45' area:

   a. 900 sq. yd.

   b. 33.33 sq. yd.

   c. 100 sq. yd.

   d. 75 sq. yd.

2. How many square yards of carpet must be ordered for a room that is 20' wide by 30' long? (This carpet only comes in a 12' roll.)

   a. 70 sq. yd.

   b. 80 sq. yd.

   c. 90 sq. yd.

   d. 100 sq. yd.

Notes

Notes

Calculate cubic yards

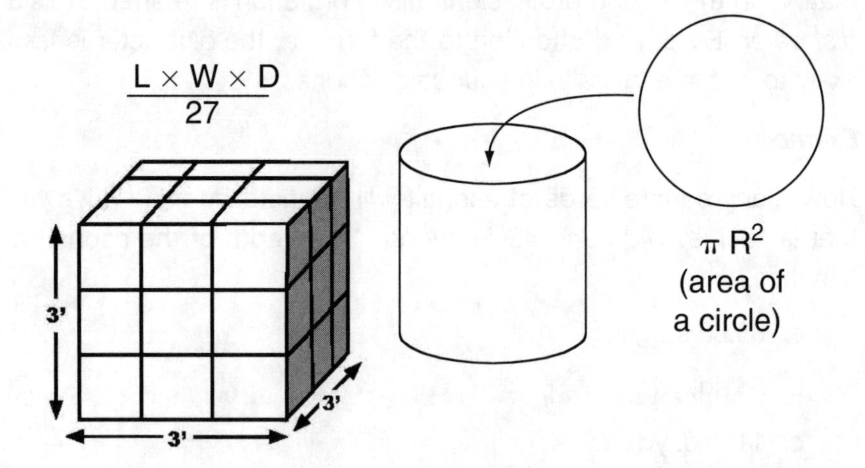

The formula to calculate cubic yards is length (in feet), times width (in feet), times depth (in feet), divided by 27. It is necessary to divide by 27 because there are 27 cubic feet in 1 cubic yard.

Cubic yards will often be referred to as *volume*. If an exam question directs you to calculate the volume of a building component, it is most likely asking for the cubic yards. However, if the choices are provided in cubic feet, do not divide by 27 (CF = L × W × D). Concrete, sand, and dirt are most commonly estimated in volume of cubic yards.

To calculate the cubic yards of a cylinder, begin by computing the area of the base. Multiply the base by the height of the cylinder and divide the total by 27.

Example:

Calculate the cubic yards of concrete needed for a 6 foot column that is 30 feet tall.

a. 20.93 cu. yd.

b. 31.4 cu. yd.

c. 565.2 cu. yd.

d. 847.8 cu. yd.

Explanation:

Begin with the formula L × W × D ÷ 27. L × W is the area of the column base computed by πR^2. R = half of the diameter, which will be 3, square the 3 to get 9. Pi × 9 equals the area, so 3.14 × 9 gives the area to be 28.26. 28.26 is the L × W portion of the CY formula. Now, multiply this by the depth of 30 feet to determine that this column has 847.8 cubic feet. Divide the cubic feet by 27 to determine the cubic yards: 847.8 ÷ 27 = 31.4. The answer is (b). (Note π is symbol for pi which always equals 3.14)

Example

How many cubic yards of concrete should be ordered for a 4"-deep drive that is 14' wide and 65' long?

a. 8.43 cu. yd.

b. 11.12 cu. yd.

c. 300.3 cu. yd.

d. 364.7 cu. yd.

Explanation:

Begin with the formula $L \times W \times D \div 27$. The length is 65, the width is 14, and the depth is 0.33 (4 divided by 12): $65 \times 14 \times 0.33 \div 27 = 11.122$. The answer is (b).

1. How many cubic yards of concrete will be needed for a slab to be poured for a parking garage if the surface area is 875 square feet and the average slab depth is 6 inches?

   a. 10.80 cu. yd.

   b. 16.20 cu. yd.

   c. 19.44 cu. yd.

   d. 437.5 cu. yd.

2. Calculate the volume for a 32' tall holding tank that is 18' in diameter.

   a. 254.34

   b. 8,138.88

   c. 301.44

   d. 354.64

3. Two hundred linear feet of footing will be poured with an average width of 14". The average depth of the footing is 16". Concrete must be ordered by the whole yard and costs $65 per unit. What is the cost of the concrete for the footings?

   a. $715.00

   b. $749.45

   c. $780.00

   d. $2,022.93

Notes

Notes

4. Based on this detail, calculate the cubic yards of sand needed to place between the slab and grade. The interior dimensions of the structure are 14' × 29'.

   a. 678 cu. yd.

   b. 25.11 cu. yd.

   c. 19.99 cu. yd.

   d. 18.65 cu. yd.

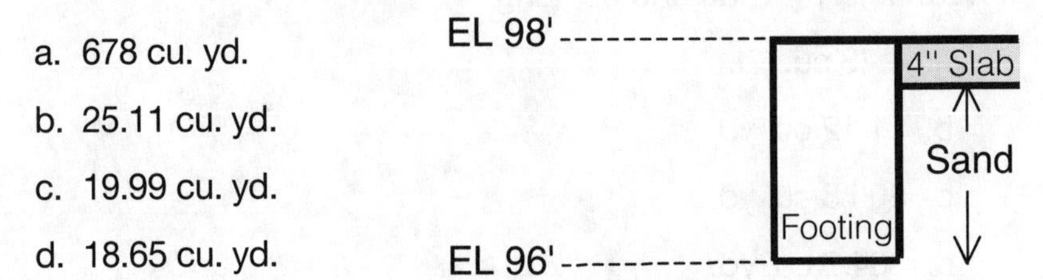

Calculating cubic yards is simple, especially when you are asked to calculate concrete. If the question is in reference to cubic yards of dirt, the swell and compaction factors need to be considered.

Swell is defined as the volume growth in soil after it is excavated. If you excavate, the soil will expand as it loosens. You cannot predict the number of dump trucks necessary to haul the excavated material away based solely on the dimensions of the area to be excavated. The volume of soil at its natural state, or bank cubic yards, will exceed the area excavated because it is no longer compacted.

The percentage of swell, or loose soil, depends on the soil type. This will be specified on the exam unless one of your approved reference books includes a chart with the information. If you are asked to calculate the cubic yards of dirt to be excavated from an area and are provided with a swell factor, simply increase the volume of the area to be excavated by the percentage provided.

Compaction rates can also be specified. When an excavation is to be backfilled, the fill must often be mechanically compacted. The compaction factors, or the difference between the fill in the truck and the final in-place compacted quantity, must be considered. If the compaction factor is 15%, divide the volume of the excavation by the reciprocal of the compaction factor. For example, if the open excavation represents 9 cubic yards and the compaction factor is 15%, divide 9 by 0.85. The volume to be delivered to fill the excavation after compaction is 10.58 cubic yards.

Example:

Based on a swell factor of 20%, how many cubic yards of dirt will be hauled away from a 12' × 22' excavation measuring 6' in depth?

a. 1,900.8 cu. yd.

b. 58.67 cu. yd.

c. 73.33 cu. yd.

d. 70.39 cu. yd.

Explanation:

Begin with the formula L × W × D ÷ 27. The length is 12, the width is 22, and the depth is 6. 12 × 22 × 6 ÷ 27 = 58.67 plus 20% = 70.399. The answer is (d).

Example:

Based on a compaction loss of 10%, how much fill is needed for an open excavation with a volume of 21 cubic yards?

a. 23.1 cu. yd.

b. 30 cu. yd.

c. 23.33 cu. yd.

d. 26.25 cu. yd.

Explanation:

21 ÷ 0.90 = 23.33. The answer is (c).

1. Based on Drawing #1, determine the quantity of soil to be hauled away from this site if it needs to be excavated 3' deep. Include an average of 2' around the entire structure. The swell factor is 15%, and the grade is considered to be level.

   a. 244.3 cu. yd.

   b. 265.5 cu. yd.

   c. 277.3 cu. yd.

   d. 280.6 cu. yd.

2. Soil with a 20% swell factor will be removed from a 4,000-square-foot excavation that is 7 feet deep. A dump truck will haul 20 cubic yards. How many dump trucks will be needed for this project?

   a. 52

   b. 63

   c. 67

   d. 71

Notes

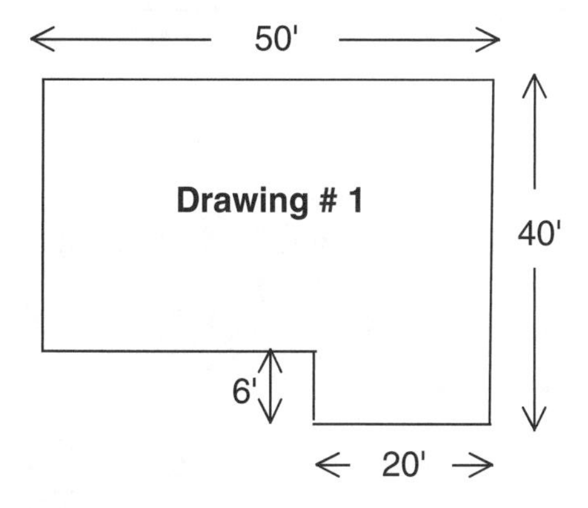

Notes

3. Considering a 10% loss of compaction, how many cubic yards of dirt will be needed to backfill an area that is 14' × 72' with a depth of 4'?

   a. 149.33 cu. yd.

   b. 164.266 cu. yd.

   c. 165.925 cu. yd.

   d. 167.854 cu. yd.

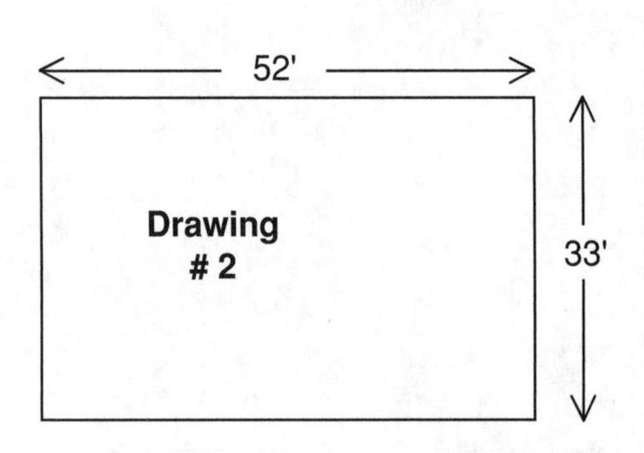

4. Based on Drawing #2, this area has been excavated 5' from the exterior of each wall. The entire area is 2' too low and must be backfilled with engineered soil that has a loss of compaction rated at 15%. How many cubic yards should be ordered?

   a. 232.33 cu. yd.

   b. 245.55 cu. yd.

   c. 254.32 cu. yd.

   d. 257.31 cu. yd.

FLOOR FRAMING TERMINOLOGY

Decking Material that forms the floor surface, usually attached directly over the floor joists

Girder A larger beam of wood or steel used as the main support for concentrated loads at points along its span

Floor joists Parallel framing members to support floors. These are the main framing members that support the floor span. Joists are usually made of engineered wood I-beams or at least 2 × 8 lumber.

Joist hangers Brackets designed to hold joist ends

Ledger strip A strip of lumber nailed to a beam, girder, or joist on which the floor joist rests for support

On center The allotted spacing between studs, joists, and rafters. This measurement is taken from the center of one member to the center of the adjacent member. On-center spacing is dictated by code and effects the joist size.

Span Distance between the structural supports in floors, ceilings, and roofs.

WALL FRAMING TERMINOLOGY

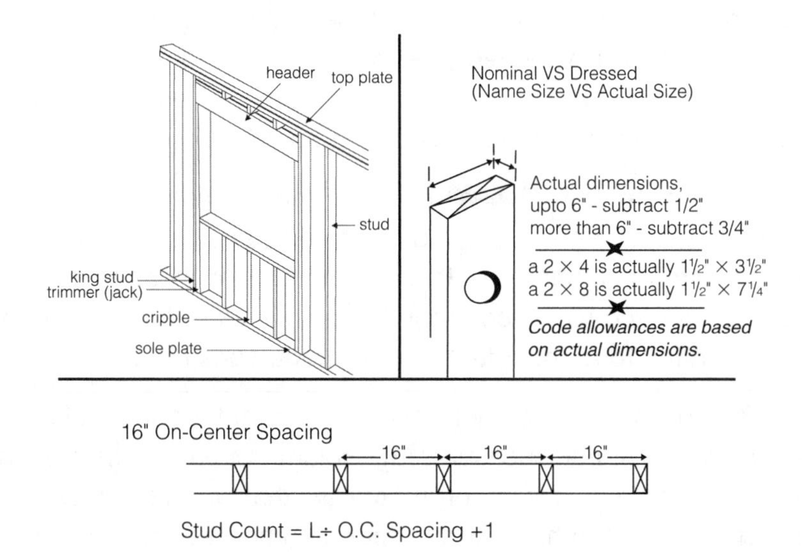

Bottom (sole) plate The bottom horizontal structural member of a stud-framed wall. The bottom plate is fastened to the floor (foundation or subfloor) with foundation bolts or proper nails.

Cripple stud The short studs placed above or below an opening. Cripples attach the header to the top plate and the sill to the sole plate.

Header A beam placed perpendicular to the wall studs above openings. The header is sized based on the span and the weight to be supported by the member.

Rough opening The framed opening to accommodate future installation of a door, window or open passageway. Measurements for placing such components are made to the center of the rough opening.

Stud A 2" × 4" or 2" × 6" vertical framing member used to construct walls and partitions.

Top plate The top horizontal framing member that caps the framed wall and makes the connection for the ceiling joists and/or rafters. Top plates are usually doubled on exterior and bearing walls and can be single members for interior nonbearing partitions.

Trimmer stud The vertical member, sometimes referred to as a jack, that supports a header.

King stud The vertical framing members on the outside of the jack or trimmer stud.

Notes

Notes

WALL FRAMING MATH

Let's begin this topic with a review of the terminology used for estimating lumber. As reviewed previously, *square feet* is the area determined by multiplying length times width. For estimating lumber, it is how much surface a milled wood product will cover.

Another common unit of measuring for construction is *linear feet*. This unit of measurement is one dimensional. It is the measurement of the framing member's length or an aggregate measurement of the entire wall length. It is common to begin a take-off (estimate) by determining the total linear feet of walls to be constructed.

Board feet, a less common unit of measurement for construction, is three dimensional and measures the volume of wood when sawn before it is kiln-dried. It is important to note that wood shrinks, more in width than in length, so the proper volume of lumber must be available before the desired finish product is planed. To estimate the quantity of board feet, multiply the quantity (number of pieces) by the nominal size of the member (in inches) by the length of the member (in feet) and divide by 12. Yes, you will break the rule of always multiplying "like numbers." This is the proper way to estimate board feet.

Nominal size is the size of the lumber before it is planed or finished. It is the "name size." The actual dimension of the framing members is not the same as the name size. The actual size of the framing members is reduced by ½" for nominal sizes up to 6" and ¾" for members above 6". For example, a 2 × 4 is actually 1½" × 3½". A 2 × 8 is actually 1½" × 7¼". Understanding this distinction is important to properly apply the building code and is critical to accurately performing math operations on the exam.

On-center spacing is an important concept to understand when estimating framing materials. If 16" on center is specified, the member will straddle the 16" incremental point across the span. The first two members are not center to center. Initially, the tape is pulled from a corner, and marks are made at the first 16-inch point and from each one that follows. Proper centering is important for the purpose of sheathing installation.

Dimensions on a set of plans are provided based on center to center of the framing. Construction materials often vary slightly in size, making it impossible to indicate actual edges of structural members. This method of measurement ensures that the plan (construction document) dimensions are achieved. If a window is to be placed, measurements are made to the center of the rough opening. The jack is installed by measuring a distance in each direction away from the center point that is equal to half of the rough opening. The size of the rough opening is always as wide as the specified component size plus 2 to 3 inches. The additional space provides for shimming the component to be installed so that it is level and plumb.

The quantity of members necessary for proper construction, such as studs, is calculated by taking the linear feet of the span and dividing it by the on-center spacing. This results in the number of spaces between the units; therefore, one additional unit must be added for the beginning (first) member. For example, if 2 × 4 studs are to be installed 16" on center in a 32' wall, the quantity of studs are calculated by taking 384" (32' × 12 to convert to "like" numbers) and dividing by 16. The result is 24. There are 24 spaces between each of the studs, add 1 and you have identified the number of studs to be installed in the wall. This same method applies to calculating the necessary number of joists or rafters.

If this 32' wall is 8' tall including a single sole plate, a double top plate, and no openings, the linear feet of lumber can easily be calculated. Begin by subtracting the thickness of the plates from the height of the wall. Since 2 × 4's are being installed, the thickness of each plate is 1½" (nominal vs. actual size). Three plates at 1½" each totals 4½". Subtract 4½" from 96" to determine that each stud will be 91½" in length.

If 25 studs are to be installed for this 32' wall, 25 × 91.5 gives us a total of 2287.5". Add the length of the plates, three plates at 32' long for a total of 96'. Now, divide 2287.5" by 12 to convert to feet. The total linear feet of lumber of studs is 190.62½. Add 96' for the plates; there are 286.62½ linear feet of lumber in this wall.

While the linear feet of lumber is rarely used in the real world, understanding the concept is critical to accurate estimating. More importantly, at this point, the concept is common on exams.

To calculate the board feet of lumber for this wall, begin by multiplying 25 × 2 × 4 × 8 and dividing by 12. This equals 133.33. This is the total number of board feet in the studs. Next, multiply 3 (the number of plates) by 2 × 4 × 32 and divide by 12. This equals 64. Add the two totals, 133.33 and 64; the total board feet for this wall is 234.66. When calculating board feet, you do not deduct for the material that will be cut, as we did when calculating the linear feet.

To "lay out a house" for the framer, it is necessary and critical to have a thorough understanding of plan reading and the ability to

Notes

Notes

work a calculator. For exam purposes, the understanding is even more critical because you will have to visualize and sketch the exercise. For example, in the real world, it would be simple to locate the position of a 3' door to be installed 10' from the edge of the room. The plans would indicate that the 10' is the center of the 3' door. An experienced carpenter would simply pull the measuring tape from the edge of the room and mark the 10' point. He would then add 2½" to the size of the door and divide the total by 2. This would provide the second and third pencil mark to be made. Half of 38½" is 19¼". The carpenter would measure 19¼" to the right of the 10' mark and 19¼" to the left of the 10' mark. These marks would indicate where the edge of the trimmer stud (jack) should be installed. The following examples indicate how this question may be presented on an exam.

Example:

The center of the rough opening for a 3' door is indicated on the plans as being 10' from the edge of the room. What is the measurement from the edge of the room to the first beginning point of the rough opening?

a. 100.75"

b. 99.25"

c. 98.75"

d. 97.25"

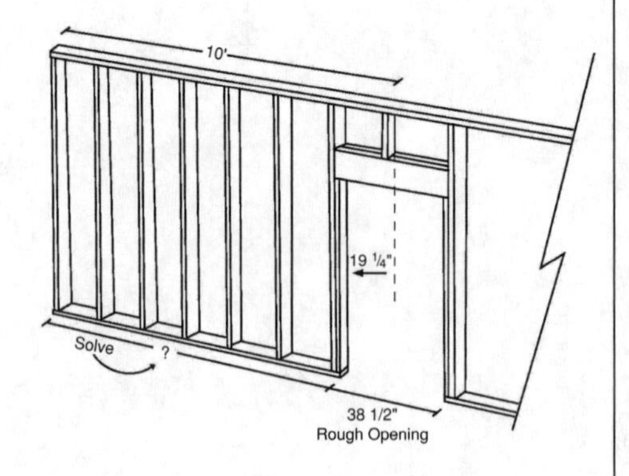

Explanation:

Begin by sketching the information as it is described. Make a dimension line from the left to right indicating 10' to the center of an opening. The opening is 38.5 inches —the 3' for the door plus 2½" for the rough opening. Divide the rough opening by 2. Draw the dimension line, indicating 19.25". Subtract 19.25" from 120": 120 − 19.25 = 100.75". This is the measurement from edge of the room to where the rough opening begins. The answer is (a).

Example:

The center of the rough opening for a 3' door is indicated on the plans as being 10' from the edge of the room. What is the measurement from the edge of the room to the first right edge of the first trimmer?

a. 100.75"

b. 99.25"

c. 98.75"

d. 97.25"

Explanation:

Begin by sketching the information as it is described. Make a dimension line from the left to right indicating 10' to the center of an opening. The question is asking for the measurement to the right edge of the first trimmer. The header rests on the trimmer; therefore, the measurement to be identified will also indicate the measurement to the beginning of the header. The measurement of the header is 3" larger than the rough opening. This is determined by adding the width of each trimmer used to hold the header in place. Each trimmer is 1½" for a total of 3". The header is 41½". Divide 41.5 by 2 to determine half of the header. Subtract 20.75 from 120". The measurement from the left to the first right edge of the trimmer is 99.25". The answer is (b).

Notes

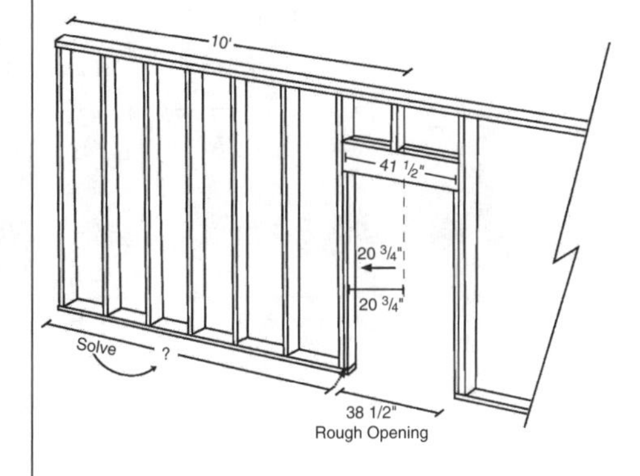

1. How many studs will be needed for a wall that is 25' long if studs are placed 16" on center?

   a. 18

   b. 19

   c. 20

   d. 21

2. Calculate the board feet of lumber in 18 floor joists. The 2 × 10 floor joists are 16' long.

   a. 27

   b. 480

   c. 1,280

   d. 5,760

Notes

3. What is the measurement to the beginning of the rough opening for a 3' door? The door will be placed with the center located 9' from the edge of the room?

   a. 86.75"

   b. 87.25"

   c. 88.75"

   d. 89.25"

4. How many floor joists will be needed for a 24' span if spaced 16" on center?

   a. 18

   b. 19

   c. 20

   d. 24

ROOF FRAMING TERMINOLOGY

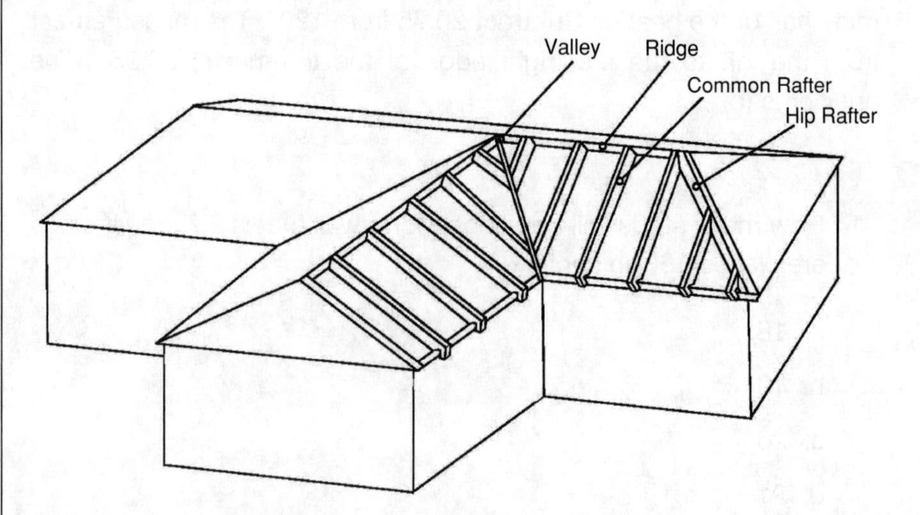

Ridgeboard The highest horizontal roof member which serves to align the rafters and tie them together at the upper end. The ridgeboard is at least one size larger than the rafters.

Common rafter A structural member that extends from the top plate to the ridge in a perpendicular orientation. Rafters often extend beyond the roof plate to form the overhang, also called the eaves, which protect the sides of the structure.

Bird's mouth The notch, or cutout, of the rafter, which allows it to rest properly on the top plate.

Hip rafter A roof member that extends diagonally from the corner of the plate to the ridge.

Valley rafter A roof member that extends from the plate to the ridge along the lines where two roofs intersect.

Jack and hip rafters These types of rafters do not extend the entire distance from the ridge to the top plate of a wall.

Cripple jack A rafter fitted between a hip rafter and a valley rafter, a cripple jack does not touch the ridgeboard or top plate.

ROOF TYPES

Gable roof Perhaps the most common, a gable roof has two slopes that meet at the center of the building. It is simple, economical, and can be used on virtually any structure.

Hip roof A hip roof has four sides or slopes running toward the center of the building. Rafters at the corners extend diagonally to meet at the ridge. Additional rafters are framed into these rafters.

Gable and Valley roof A roof on which two gable roofs intersect.

Hip and Valley roof A roof on which two hip roofs intersect.

Mansard roof A roof that has four sloping sides, each of which as a double slope. As compared with a gable roof, this design provides more available space in the per level of the building.

Gambrel roofing A variation on the gable roof in which each side has a break near the ridge. This style of roof will provide more available space in the upper level.

Shed roof Also referred to as a lean-to roof, the shed roof is a flat, sloped construction. It is common on high-ceiling contemporary construction and is often used on additions.

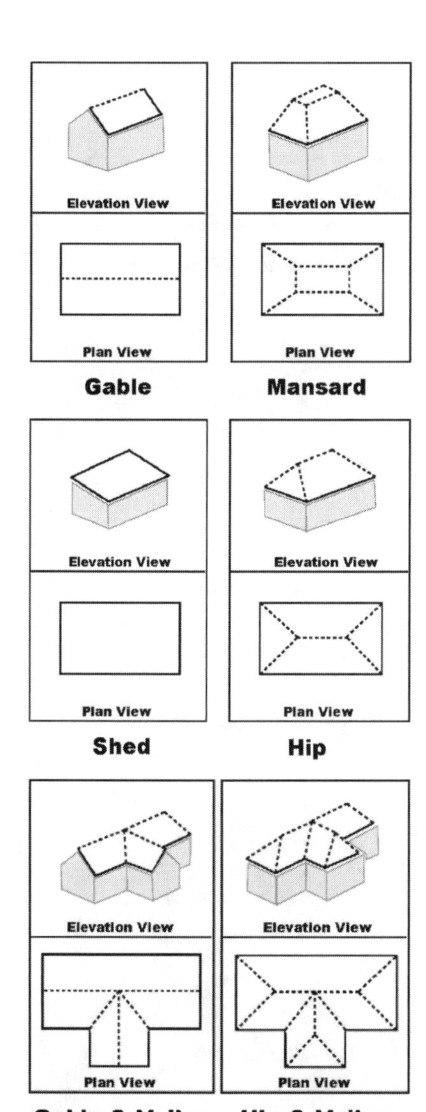

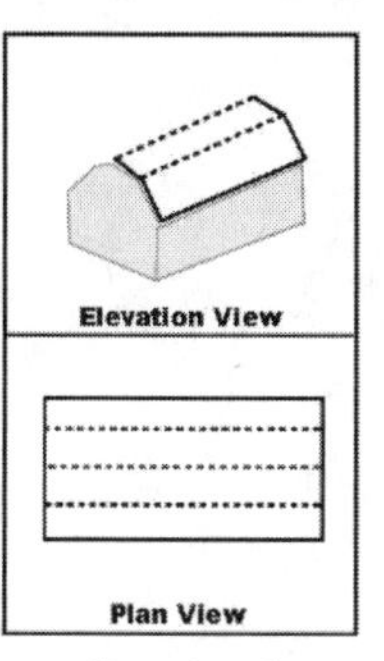

Gambrel

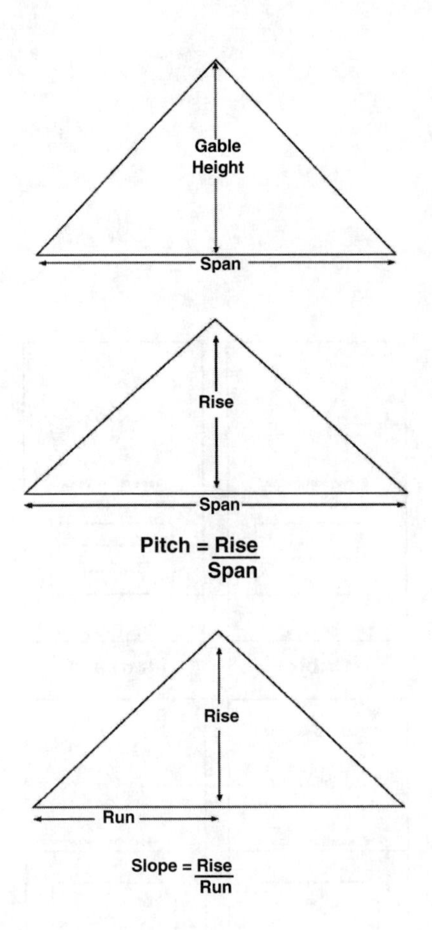

ROOFING MATH

To begin the discussion of roofing math, there are several terms that must be clearly understood. It is important to understand that the term *span* refers to the horizontal distance from the outside of one exterior wall to the outside of the opposite wall. *Run* is the horizontal distance from the outside of the top plate to the center line of the ridgeboard, usually equal to half the span.

Rise is the total height of the rafter from the top plate to the ridge. The rise is typically stated in inches per foot of run.

Pitch is the angle or degree of slope of the roof in relation to the span. It is a ratio of gable height over the full span of the roof. Pitch is expressed as a fraction. For example, if the total rise is 6' and the span is 24', the pitch would be 6 in 24 which reduces to a ¼ pitch.

Slope is expressed as the relationship of rise to run. It is stated as a unit of rise to horizontal units of 12. A roof that has a rise of 5" for each foot of run is said to have a 5 in 12 slope.

It is common for a seasoned contractor to refer to a 5 in 12 slope as a 5⁄12 pitch. No one corrects him and asks, "are you sure you don't mean a 5 in 12 slope?" Notice, the only words that change between the two statements are *pitch* and *slope*. These terms are used interchangeably, and both uses are accepted as correct. In fact, it is so common that even the codebooks refer to a slope as a pitch. Let's look at some examples to make sure that this concept is understood.

If a building is 24 feet wide and has a gable height of 8 feet, by placing the height over the building span, the pitch will be 8 over 24. This fraction reduces to ⅓.

If you have forgotten how to reduce a fraction, simply divide the numerator (top number of the fraction) and the denominator (bottom number) by their greatest common factor or devisor. If you have properly reduced the fraction, only the number 1 can be divided evenly into both the final numerator and denominator.

The slope of this roof would be 8 in 12, because a slope is expressed in a ratio of rise over run. Half of the 24' span would be a 12' run. The rise of 8 and the run of 12 is expressed in a ratio format of 8 in 12.

Let's try another example: If the building is 32' wide with a gable height of 8', by placing the numbers in a ratio of gable height over the span of the building, you will have 8 over 32. Now, reduce the fraction by taking the greatest common factor and dividing it into the numerator and the denominator. The greatest common factor is 8, and 8 will go into 8 one time, and 8, will go into 32 four times, with the result being that this roof has a one-fourth pitch.

If you need to determine the slope of this roof, converting the $\frac{1}{4}$ pitch to a slope is a little more difficult. Roof slopes are always provided with a ratio of the specified units of rise to 12 units of horizontal run. The run is always indicated in units of 12. This is because we measure in units of 12". When asked, a contractor usually provides measurements in feet. Rarely will he say "the building is 480" wide."

So, what is the slope of a roof 32' wide with an 8' gable height? Many would quickly say 8 over 16. Yes, but this would not be one of the choices on the exam. The choices would be in a ratio of rise to run with the run being 12. So, let's set it up to solve for x (or the correct unit of rise).

Place x over 12, and beside it put 8 over 16. We will now cross multiply and divide by the reciprocal to solve for x: $12 \times 8 = 96$. Divide 96 by 16, which equals 6. The slope of this roof is 6 in 12.

Step 1: $\frac{x}{12} \quad \frac{8}{16}$

Step 2: $\frac{x}{12} \nearrow \frac{8}{16}$

(cross multiply)

$12 \times 8 = 96$

Step 3: Divide by reciprocal:

$96 \div 16 = 6$

Example:

Determine the slope of a roof in units of vertical rise to 12 units of horizontal run when the span of the roof is 30' and the height of the gable is 10'. By placing the rise over the run, you have 10 over 15. However, we need it in units of rise to units of horizontal run. Again, we will solve

Notes

Notes

for x over 12 and beside it write the fraction 10 over 15. We will cross multiply, so 12 times 10 will equal 120. Now, divide by the reciprocal of 15 to get 8. The slope of this roof is 8 in 12.

Step 1: $\frac{x}{12} \quad \frac{10}{15}$

Step 2: $\frac{x}{12} \quad \frac{10}{15}$

(cross multiply)

$12 \times 10 = 120$

Step 3: Divide by reciprocal:
$120 \div 15 = 8$

If provided a pitch in its reduced fraction such as ⅓, follow the same steps. The slope will be 1 over 1.5 (If 3 equals the span, run is half hence the 1.5.) To convert this to a slope, place x over 12 beside the fraction 1 over 1.5. Cross multiply 12 times 1 and divide by the reciprocal of 1.5. The result is 8. A roof with a ⅓ pitch will have an 8 in 12 slope.

Step 1: $\frac{x}{12} \quad \frac{1}{1.5}$

Step 2: $\frac{x}{12} \nearrow \frac{1}{1.5}$

(cross multiply)

$12 \times 1 = 12$

Step 3: Divide by reciprocal:
$12 \div 1.5 = 8$

Questions on this topic generally use whole numbers and reduced pitches. It may be to your advantage to make a chart with three columns. The first column will be labeled "slope," the second column will be labeled "pitch," and the third column will be labeled "reduced pitch." First, fill in the slopes beginning with 2 in 12. Continue down to 8 in 12. In the second column, convert the pitch to a slope by multiplying12 by 2 and placing the vertical rise on top (see example). You have converted the run to a span and now have a ratio of rise over span. In the last column, reduce each of the fractions.

A typical exam question is What is the vertical rise of a roof that has a ⅓ pitch?" Using your chart, locate the ⅓ pitch, follow it to the column labeled "slope," and identify the vertical rise to be 8.

Run

Rise

$\frac{\text{Rise}}{2 \text{ x Run}}$

Slope	Pitch	Reduced Pitch
2:12	2/24	1/12
3:12	3/24	1/8
4:12		
5:12		
6:12		
7:12		
8:12		
9:12		
10:12		

Example:

What is the slope of a gable roof with a 36' span? The height of the gable is 12'.

a. 2:12

b. 4:12

c. 5:12

d. 8:12

Explanation:

Place *x* over 12, and beside it write 12 over 36. Cross multiply: 12 × 12 = 144. Divide by 18 (half of the 36' span) = 8. The answer is (d).

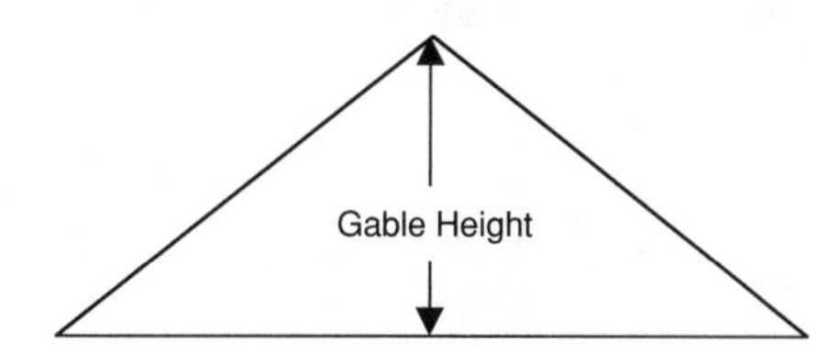

Determining the gable height of a structure is simple as long as the roof slope and span are provided. If the span of a building is 12' and the slope is 8 in 12, take half of the span and multiply it by 8". This formula is confusing to some people because it seems to be multiplying feet by inches (breaking the rule to always multiply "like numbers"). Do not think of the 6 as feet. Consider the 6 to be units and tell yourself that for each of the 6 units (units in 12-inch increments), the roof rises 8". (See diagram below)

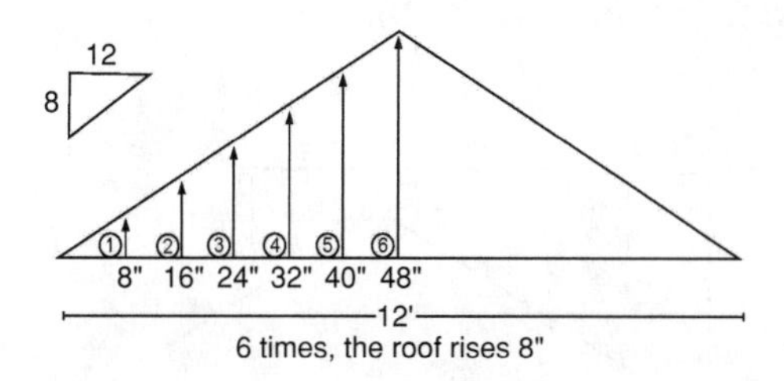

For a building with a span of 24' and a roof sloped 4 in 12, first determine the measurement to the ridge. If the building is 24' wide, the ridge will be located in the center. Half of 24 is 12. Now, multiply 12 by the vertical rise for a total of 48 inches. Divide 48 by 12, and the gable height is determined to be 4 feet.

Understanding this concept allows you to easily calculate the area (square feet) of the gable. Once it is calculated, the area can be multiplied or divided by the appropriate factor to estimate siding, brick, sheathing, and similar building components.

Notes

Notes

If you need to estimate the quantity of brick (based on a factor of 7) to be installed in the gable area of a building that is 30' wide with an 8 in 12 slope roof, begin by converting the 30' span to a run. Divide 30 by 2. The run of the building is 15'. Next, multiply 15 by the vertical rise of 8. This equals 120. The height of the gable is 120" or 10'. Determine the area of the gable by taking the length of the run and multiplying it by the height of the gable. The area of the gable is 150 square feet. Since the actual area is only half of the rectangle that you calculated, it will account for each portion of the gable area. Take the square feet of the gable, and multiply it by 7 to determine that 1050 brick will be needed for this area.

Example:

Calculate the square feet of a gable for a building that is 28' wide and built with a 4:12 slope.

a. 32.69 sq. ft.

b. 65.38 sq. ft.

c. 78.36 sq. ft.

d. 95.25 sq. ft.

Explanation:

Identify the center point of the span, which is the highest point of the gable: $28 \div 2 = 14$. This roof rises 4" for each foot. Multiply 14 by 4 to determine the gable height: $14 \times 4 = 56$. Convert 56 inches to feet: $56 \div 12 = 4.67'$. Multiply 14 by 4.67 for a total of 65.38 square feet. If calculating the area of a triangle, divide by 2. It is not necessary to divide by 2 in this case because you have two triangles, one on each side of the ridge. The answer is (b).

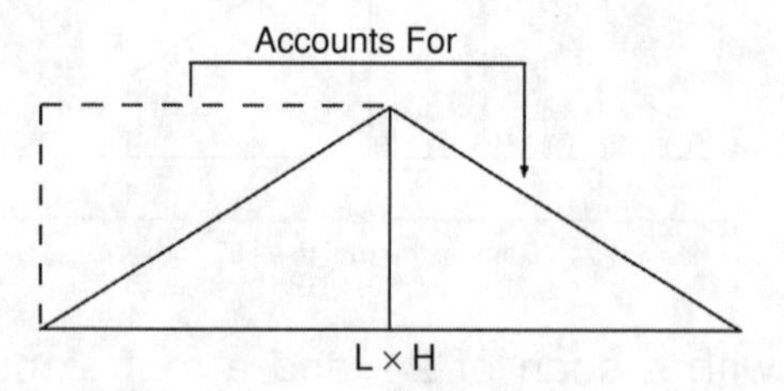

1. What is the gable height for a building that is 28' wide? The roof slope is 6 in 12.

a. 7 ft.

b. 8 ft.

c. 9 ft.

d. 10 ft.

Notes

2. What is the slope of a roof with a ¼ pitch?

   a. 4:12

   b. 5:12

   c. 6:12

   d. 8:12

3. What is the run of a roof that has a 29' span?

   a. 13 ft.

   b. 13.5 ft.

   c. 14 ft.

   d. 14.5 ft.

4. Calculate the area of a gable if the slope of the roof is 8:12. The width of the building is 32'.

   a. 256.00

   b. 170.72

   c. 225.80

   d. 154.75

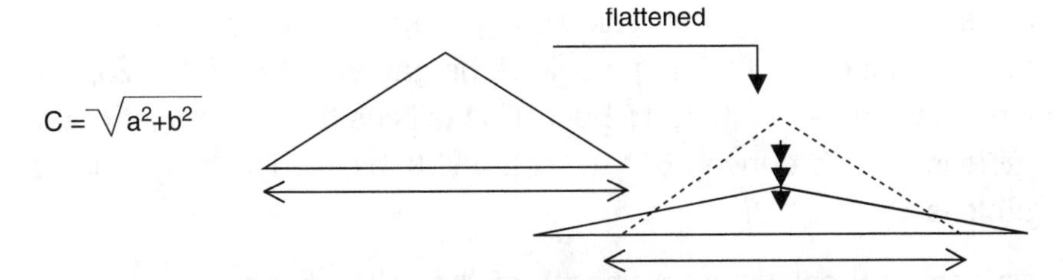

The next concept is critical for many areas of roofing math, including estimating rafter lengths, bundles of shingles, and rolls of felt. For explanatory purposes, suppose we are working with a gable roof that is 24' wide with a 4 in 12 slope. Imagine that the ridgeboard is hinged to allow the roof to lay flat. If this roof were laid flat, it would not be 24' wide. Obviously, it would be wider. The question is, how much wider?

To determine the measurement of a building if we could lay it flat, we must apply the Pythagorean theorem. This theory states that when we have a right triangle, one that is formed with a 90-degree angle at the base, we can determine the hypotenuse as long as we

Notes

know the two remaining measurements. Speaking in construction terms rather than mathematical jargon like triangles and hypotenuses, the right triangle is one half of our 24' wide building. The hypotenuse is the rafter on the building that connects the base to the ridge.

The formula states that *C*, the rafter, is equal to the square root of *A* squared plus *B* squared. *A* is the run of the building, and *B* is the rise of the building. In this case, *A* is 12, and *B* is 4. Plug the numbers into the formula. Begin by multiplying 12 by 12. This equals 144. Next, multiply 4 by 4. This equals 16. Add the two together; 144 plus 16 equals 160. Find the square root symbol on your calculator, and press it. The numbers returned are 12.6491. This tells us the length of the rafter for this building. Double this number, and the dimension of the roof is 25.2982. If we were answering a question on the exam, it would likely have asked for the rafter length to be ordered. Since lumber is ordered in 2' increments, the correct choice for this question is 14'.

The 24' wide building with a sloped roof of 4 in 12 is the most simple of this type of mathematical operation. If the width of the building is 30' and the slope remains 4 in 12, another step must be included in solving the equation. It is important to realize why this is so. In the example of the 24' wide building, the height of the gable happened to be the same as the rise in the 4 in 12 slope. If the building were 30' wide with a 4 in 12 slope, the gable height would be different. The *B* in the formula stated in the Pythagorean theorem as we used it represents the gable height. So, if the building is 30' wide, begin by calculating the height of the gable. The center point of the building (peak of the gable) is 15'. If the slope is 4 in 12, we must multiply 15 by 4. This tells us that the height of the gable is 60". To convert 60" to feet, divide by 12. The height of the gable is 5'.

We can now calculate the length of the rafter (hypotenuse) of the building (half of which represents a right triangle). *A* is 15'. Fifteen squared equals 225. *B* is 5'. Five squared equals 25. Add 225 and 25, for a total of 250. Enter 250 in the calculator, and press the square root symbol $\sqrt{\ }$. The calculator returns 15.8113. The material to be ordered for the rafters in this building will be 16' in length.

Example:

A gable roof is built with an 8:12 slope. The building is 32' wide. If the two rafters that meet at the ridge are laid end to end, what is the total linear feet?

a. 19.23 linear feet

b. 34.25 linear feet

c. 36.84 linear feet

d. 38.46 linear feet

Explanation:

First, determine the gable height, take half of 32, which identifies the location of the highest point of the gable, and multiply 16 by 8 (this roof rises 8" for each foot): $16 \times 8 = 128$". Convert to feet: $128 \div 12 = 10.67$' (gable height). Now, plug the numbers into the Pythagorean theorem. $A = 16$, $B = 10.67$, $16 \times 16 = 256$, $10.67 \times 10.67 = 113.85$. Add these together: $256 + 113.85 = 369.85$. Press the square root symbol to get 19.23. $19.23 \times 2 = 38.46$. The two rafters end to end are 38.46'.

If the roof multiplier for a 4 in 12 slope roof had been available for the 30' wide building, the calculation would have been much simpler. If we were calculating the area of the roof, we would have multiplied the 30' by the multiplier. If we were calculating the length of a rafter, we would have divided the 30 by 2 to determine the run and multiplied the run by the factor, which would give us the rafter length. The step for determining the gable height would have been eliminated.

The roof multiplier for a 4 in 12 slope roof is 1.054. How is this calculated? Simply, plug the numbers into the Pythagorean theorem, and divide the answer by 12. This indicates the degree that the rafter (hypotenuse) rises per foot of run.

Let's calculate the roof factor for a 6 in 12 sloped roof. Plug the numbers into the formula for Pythagorean theorem, *C* equals the square root of *A* squared plus *B* squared. Twelve squared equals 144; 6 squared equals 36; 144 plus 36 equals 180. Press the square root symbol on the calculator $\sqrt{\ }$, and 13.4164 is returned. Finally, divide this by 12. Carry the factor out to the third digit. The roof multiplier, or factor, is 1.118.

Notes

Notes

It is not uncommon for the exam to provide the roof multiplier in the instructions for the related math question. Of course, if you do not realize how to use the factor, it does you no good. If the factor is not provided, it can be easily be calculated.

Calculate the factors for each of the slopes and fill in the table:

Slope	Equation	Multiplier
2:12	2 × 2 = 4, 12 × 12 = 144, 4 + 144 = 148, press √ = 12.1655, divide by 12 = 1.013.	**1.013**
3:12		
4:12		
5:12		
6:12		
8:12		

5. Calculate the length of a common rafter for a 20' × 50' gable roof. The roof slope is 4:12, and the building has a 1' overhang.

    a. 10 ft.

    b. 12 ft.

    c. 13 ft.

    d. 22 ft.

6. If 6:12 gable roof is 30' wide with a 1' overhang, what is the measurement to use in calculating the roof area?

    a. 33.54

    b. 34.658

    c. 35.776

    d. 36.516

7. Using the Pythagorean theorem, what is the rafter length for a building that is 27' wide and measures 6' at the gable height?

   a. 12.56 ft.

   b. 14.77 ft.

   c. 15.23 ft.

   d. 16.85 ft.

Roof area can easily be calculated using the multiplier, allowing us to compute the bundles of shingles, rolls of felt, or pieces of sheathing needed for the roof. You can simply calculate the square feet of the home and increase it by the roof multiplier. You may be asking, "What if it is a hip roof as opposed to a gable roof?" Believe it or not, it matters so little that it is acceptable practice to estimate both using this method.

Can you simply take the square feet provided as square feet for the roof area? You could, but if the structure has an overhang, it has not been calculated in the finished square feet. The overhang is the portion of the roof that extends beyond the wall of the building. There is no standard overhang. In fact, overhangs vary based on different roof pitches, architectural styles, and personal preferences. A structure can have multiple overhangs and be different for gable ends than what is specified for the eaves.

The simple way to begin calculating the area of the roof is to sketch the plan view of the structure. This is simply a two-dimensional sketch of the footprint of the building as if you were looking down on top of it. Next, you should place a dotted line around the perimeter of your sketch to illustrate the overhang. The area of the roof should be calculated using the new dimensions. This will provide the area of the roof if it were flat. Once this calculation is achieved, you simply multiply this number by the roof multiplier to account for the rise created by the pitch.

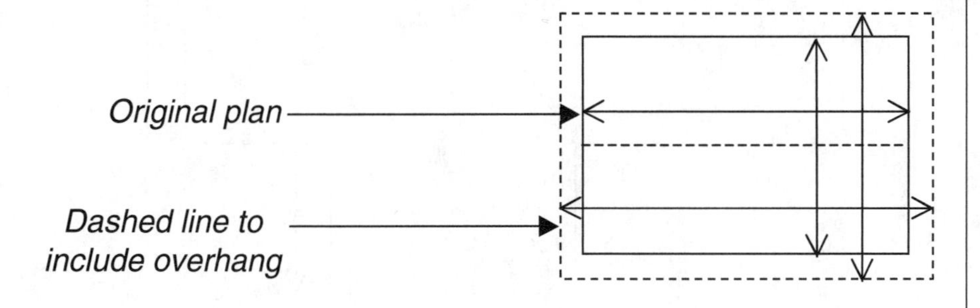

Notes

Notes

Let's try a simple roof. Calculate the roof area of a building that is 30' wide, 45' long, with a roof slope of 5:12. The overhang for the building is 1 foot. Begin by sketching the 30 × 45 foot rectangle. Next, draw a dotted line around the perimeter sketch to indicate the overhang. Now, draw a dimension line to connect the dashed lines on each side, and write the new dimensions to include the original measure plus the overhang.

The new dimension will be 30' plus 1' of overhang on each end, for a total of 32'. The dimension line going in the opposite direction will be 45' plus 1' of overhang on each end, for a total of 47'. Multiply 47 by 32 to determine the square feet of the flat area of this structure including the overhangs. Forty-seven feet times 32' equals 1,504 square feet. Finally, multiply the 1,504 square feet by 1.083. This is the multiplier for a 5 in 12 sloped roof. The total square feet of the roof for this building is 1,628 square feet.

Let's try another example. Calculate the area of a hip roof that is 75' in length, 27' in width, and has an 18" overhang. The slope of the roof is 6 in 12. Sketch the footprint of the building. Next, draw a dashed line around the perimeter of the building sketch to indicate the overhang. Add dimension lines to connect the dashed lines that are parallel to one another. The new dimensions will be 78' × 30'. Multiply 78 × 30; the area of the roof is 2,340 square feet. Multiply 2,340 square feet × 1.118, the multiplier for a 6:12 slope, to determine that the total area of the roof is 2,616 square feet.

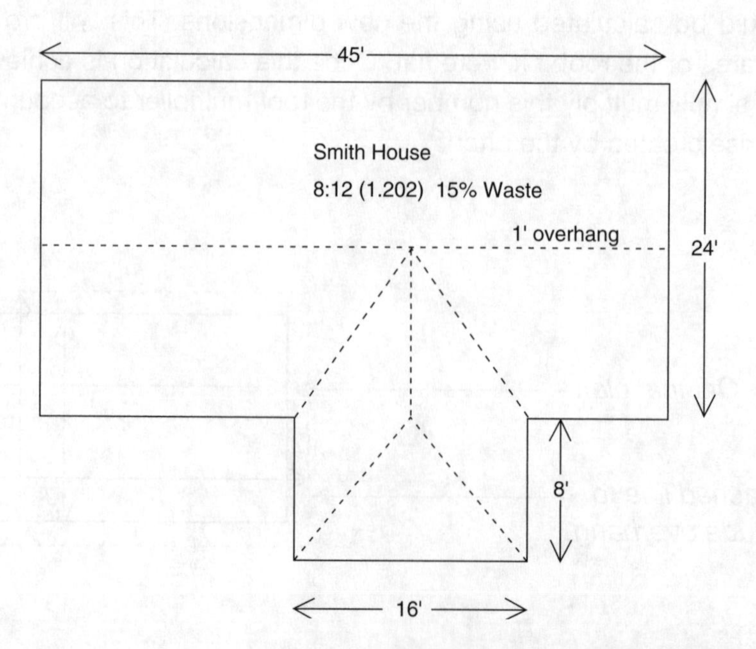

Now, one that is a little more difficult (Smith House on page 44). The building in the diagram provides all of the information necessary for calculating the area of the roof. It provides the roof slope as well as the roof multiplier. The waste factor for this plan is indicated, and the overhang is noted. Follow these steps:

In Step 1, the dashed line is added to indicate the overhang, In Step 2, the new dimension lines were added to show the overhang. Step 3 divided the plan into manageable sections. The larger of the two sections measures 47' × 26'. The smaller section is 18' × 8'. Notice that the 8-foot dimension line for the small section stays the same because of the one foot included in the larger section (indicated by the heavy line).

The larger section is 1,222 square feet, and the smaller section is 144 square feet. Added together, the total for the roof area is 1,366 square feet. To account for the pitch, multiply 1,366 × 1.202, which equals 1,641.93. Add 15% to consider the waste. The total area of roofing for this plan is 1888.22 square feet.

If you were asked to calculate the bundles of 15#, three-tab shingles for this plan, you would divide the total area by 100. This is to convert the roof area to squares. There is 100 square feet per roofing square. Shingles are packaged such that three bundles will cover 100 square feet. By multiplying the squares by 3, you determine that this plan will require 57 bundles of shingles (1,888.22 ÷ 100 = 18.88 × 3 = 56.64, which rounds to 57 bundles).

Questions on the exam typically include the roof multiplier, considerations for waste, the number of bundles required per square, yield for felt, and a statement to disregard starter shingles and ridge caps.

Notes

Step 1

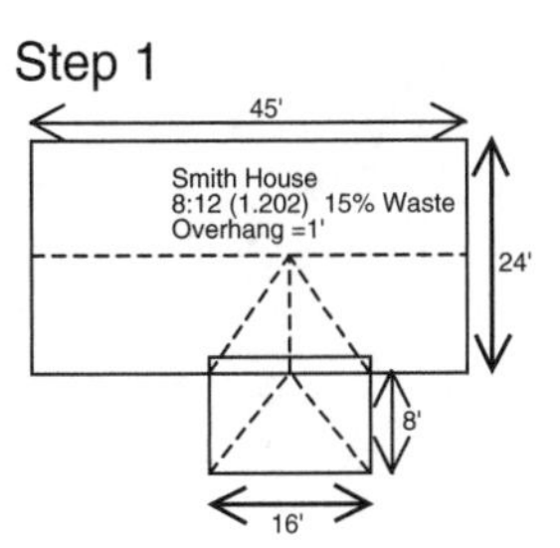

Sketch dash line to indicate overhang.

Step 2

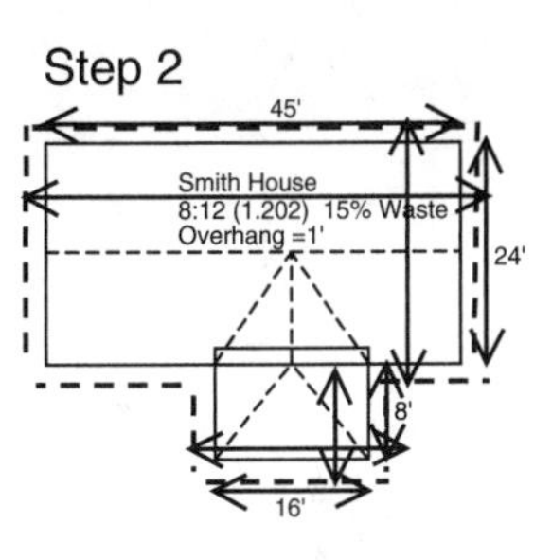

Add dimension lines to include overhang.

Step 3

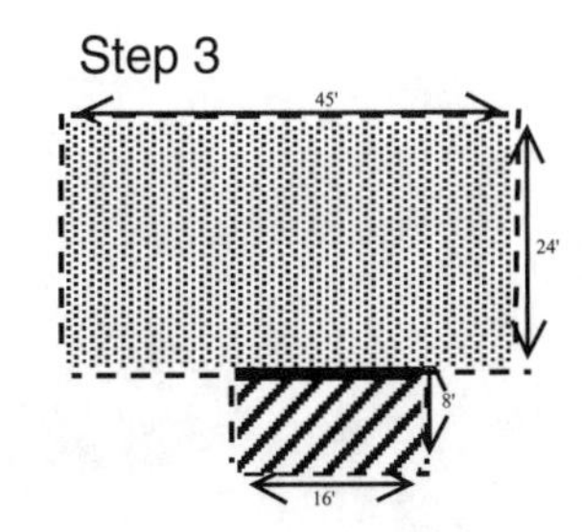

Divide the plan into sections.

Notes

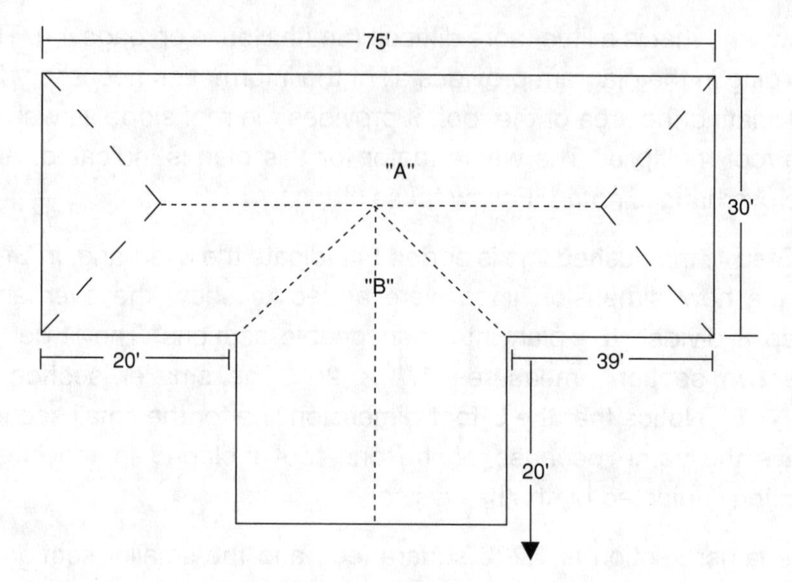

8. Calculate the square feet of this building. *Note:* Roof slope is 4:12, overhang is 1', waste factor is 10%.

   a. 2,570 square feet

   b. 2,824 square feet

   c. 2,976 square feet

   d. 3,274 square feet

9. What is the length of the common rafters to be ordered for section *B* if the roof slope is 6:12? *Note:* The overhang is 18".

   a. 9 ft.

   b. 10 ft.

   c. 11 ft.

   d. 12 ft.

10. Three-tab, 15# shingles are to be used for this residential building (3 bundles per square). The slope of the roof is 5:12, and the overhang is 1'. Waste should be based on 5%. How many bundles are needed?

    a. 72

    b. 81

    c. 91

    d. 97

PART TWO

Administration

Questions are based on the Building Code.

The following code quiz is based on topics that concern the building administration. Exam candidates preparing for an exam based on the *International Residential Code* and the *International Building Code* will find the answers in Chapter 1.

The code quizzes are written to familiarize you, the exam candidate, with the reference book. If you feel that you know the answer to the question, the correct answer is less important than your ability to identify the corresponding code location. Remember, these quizzes are representative of the subject matter covered on exams but do not represent actual questions. Do not attempt to memorize the answers. It is much more effective to understand how to locate the subject matter in the approved references.

Questions in this quiz cover the following subject matter:

- Permits
- Certificates of occupancy
- Demolition
- Temporary permits
- Building officials
- Construction documents

Study Tip

Plans or Blueprints will be referred to as Construction Documents *in your Code Book.*

1. Which of the following statements is *true* regarding site plans submitted for demolition projects?

   a. Site plans are not required for structures to be removed.

   b. Site plans must show the construction to be demolished.

   c. Site plans must show the size and location of existing construction that will remain.

   d. Both b and c

2. A permit that has been issued is deemed void if work does not begin within ___ days of issuance.

   a. 30

   b. 60

   c. 90

   d. 180

3. In addition to one set of approved plans being retained by the building official, an approved copy must always be available:

   a. At the job site

   b. At the contractor's office

   c. When the inspector requests to see a copy

   d. At the county clerk's office

4. The building permit must be:

   a. Displayed in the contractor's office

   b. Posted on the job site

   c. Provided to the inspector when requested

   d. Posted in the building official's office

5. Which of the following construction projects would *not* require a building permit?

   a. A 400 square foot, two-story building

   b. A one story, 200 square foot home attached to an existing two-car garage

   c. A one-story detached accessory structure with 120 square foot

   d. None of the above

6. Who has the authority to approve the use of temporary power?

   a. Electrical contractor

   b. General contractor

   c. Home owner

   d. Building official

7. Which of the following is not found on a Certificate of Occupancy?

   a. If an automatic sprinkler is provided

   b. Name and address of the owner

   c. Name of the building official

   d. Contractor's license number

8. Which of the following *must* be included on a stop work order?

   a. Contractor's name and license number

   b. Name of person doing the work

   c. Conditions under which work will be permitted to resume

   d. Number of days the order will be in place

Study Tip

Any topic that is related to Administrative Issues from an Inspector's perspective will be discussed in Chapter 1 of your Code Book.

Study Tip

As you locate answers to each of the questions in your Code Book, make sure to highlight them as you go.

9. One set of approved construction documents must be retained by the building official for not less than how many days from the date of the completion?

   a. 90

   b. 180

   c. 30

   d. 120

10. One set of approved construction documents shall be retained for not less than 180 days by which of the following?

    a. Contractor

    b. Home owner

    c. Building official

    d. None of the above

PART THREE

Design and Planning

Questions are based on the Building Code.

The following code quiz is based on topics primarily effecting the design of a building. Exam candidates preparing for an exam based on the *International Residential Code* will find the answers in Chapter 3. Candidates using the *International Building Code* will find the answers in various chapters.

As stated previously, the code quizzes are written to familiarize you with the reference book. Do not attempt to rely on your field knowledge to pass the exam. It is imperative for you to become familiar with using the code book. These quizzes will prepare you for the exam in the most efficient manner. Take the time to locate the answer to each question using the code book.

Questions in this quiz cover the following subject matter:

- Ceiling heights
- Lighting and ventilation
- Means of egress
- Rescue opening
- Guardrails
- Stairs
- Sanitation
- Live Loads/Dead Loads

Notes

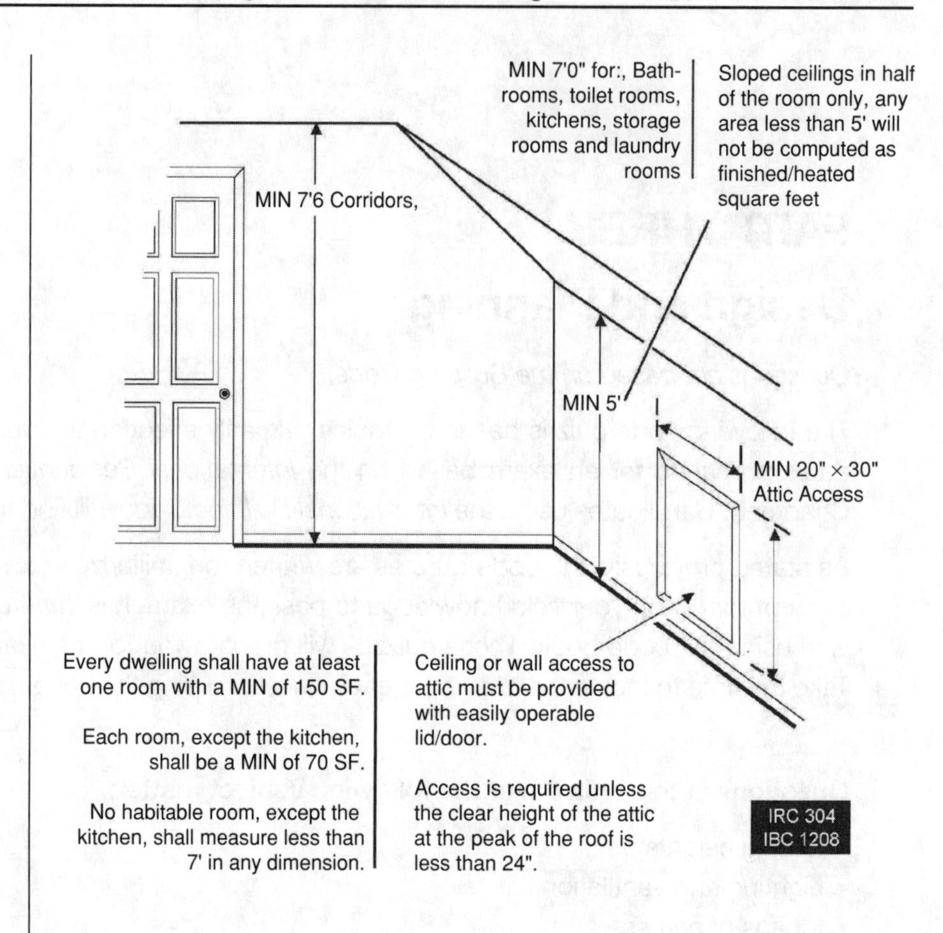
MIN 7'0" for:, Bathrooms, toilet rooms, kitchens, storage rooms and laundry rooms
Sloped ceilings in half of the room only, any area less than 5' will not be computed as finished/heated square feet
MIN 7'6 Corridors,
MIN 5'
MIN 20" × 30" Attic Access
Every dwelling shall have at least one room with a MIN of 150 SF.
Each room, except the kitchen, shall be a MIN of 70 SF.
No habitable room, except the kitchen, shall measure less than 7' in any dimension.
Ceiling or wall access to attic must be provided with easily operable lid/door.
Access is required unless the clear height of the attic at the peak of the roof is less than 24".
IRC 304
IBC 1208

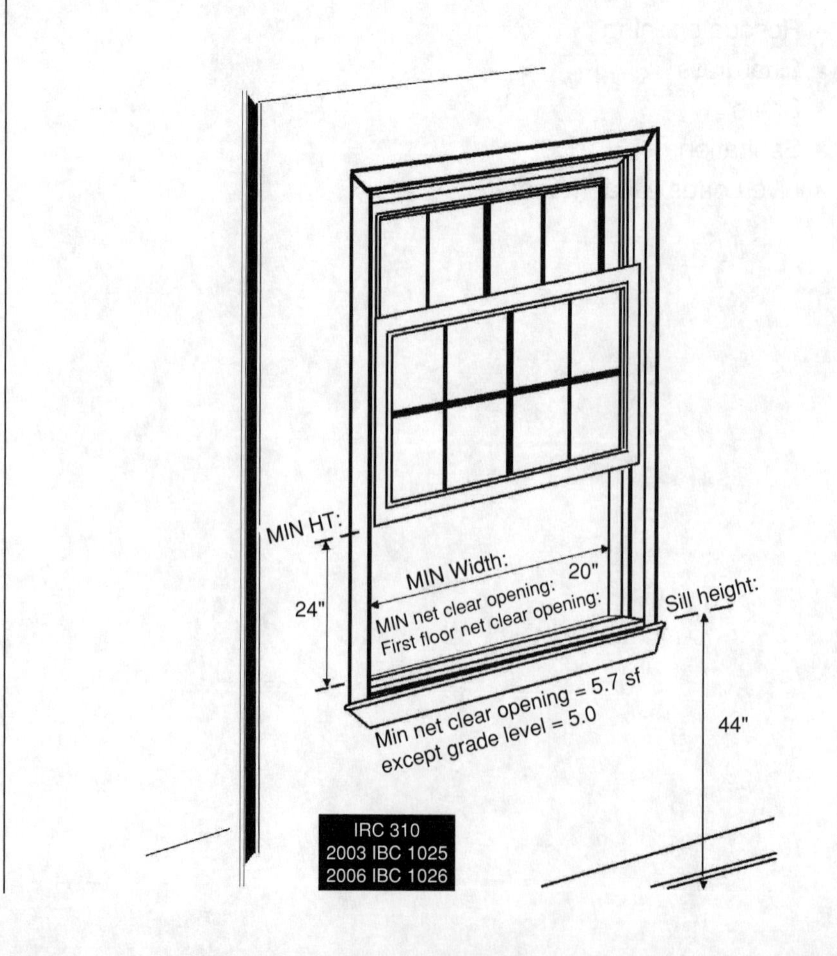
MIN HT:
24"
MIN Width: 20"
MIN net clear opening:
First floor net clear opening:
Sill height:
44"
Min net clear opening = 5.7 sf
except grade level = 5.0
IRC 310
2003 IBC 1025
2006 IBC 1026

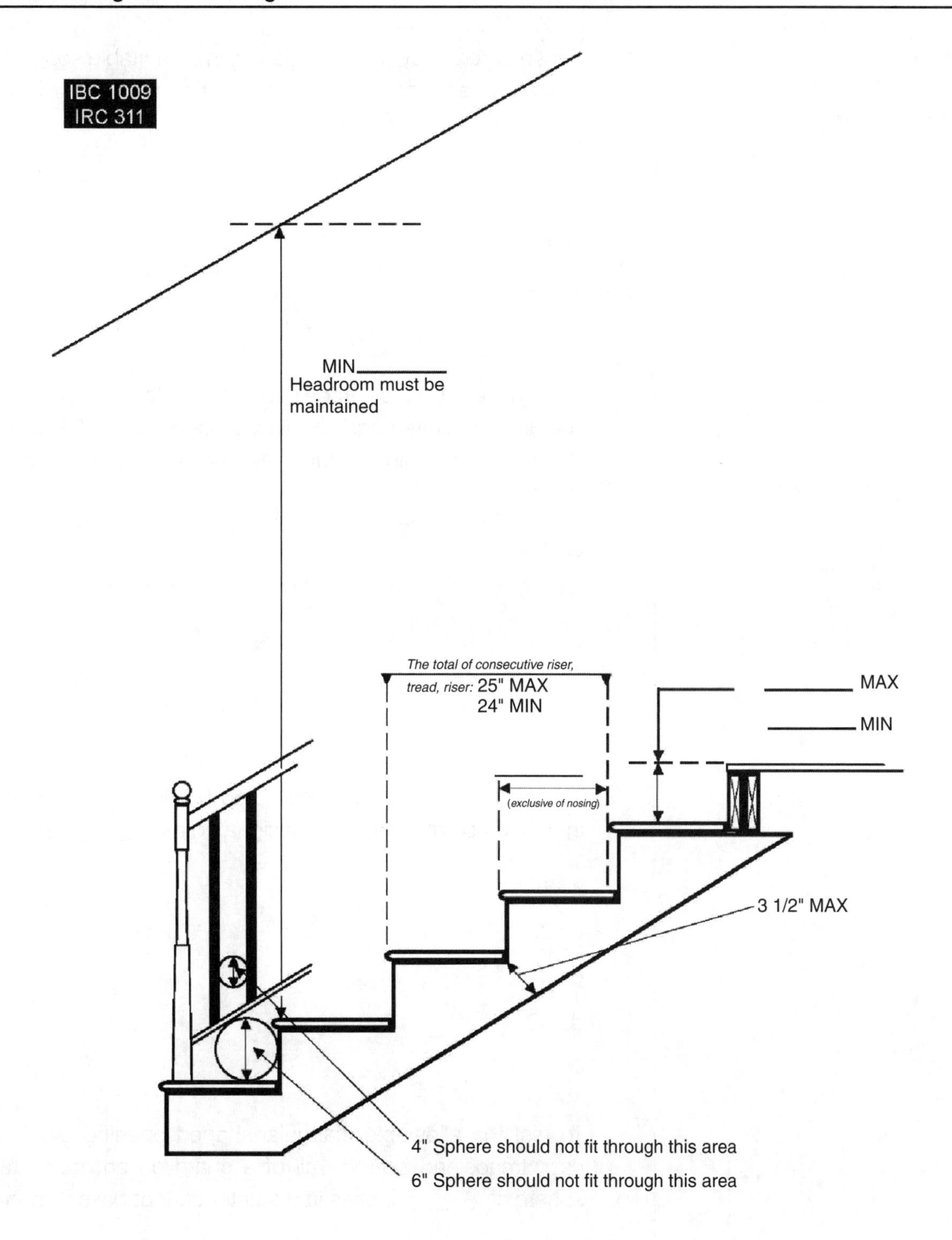
IBC 1009
IRC 311
MIN
Headroom must be maintained
The total of consecutive riser, tread, riser: 25" MAX
24" MIN
MAX
MIN
(exclusive of nosing)
3 1/2" MAX
4" Sphere should not fit through this area
6" Sphere should not fit through this area

Study Tip

Emergency Escape can also be referred to as Exit *or* Egress.

1. For an emergency escape opening in a habitable space, the minimum net clear opening width is specified as ______ inches.

   a. 20

   b. 22

   c. 24

   d. 28

2. Open access from a garage into a room intended for sleeping purposes is prohibited and should be closed. If a solid wood door is installed, the minimum thickness of the door must be ______ inches.

   a. 1¼

   b. 1½

   c. 1⅜

   d. 2

3. The minimum uniformly distributed live load for a room with an intended use of sleeping is designated as ______ psf.

   a. 10

   b. 20

   c. 30

   d. 50

4. To test the allowable triangular-shaped opening created by the riser, tread, and bottom rail of a guard on an open stairway, a sphere of ______ inches in diameter cannot pass through.

   a. 4

   b. 6

   c. 8

   d. 10

5. For ramps intended for use other than specifically for means of egress, the maximum allowable slope is:

   a. 8%

   b. 10%

   c. 10.5%

   d. 12.5%

6. Untreated wood joists should not be installed closer than ______ inches.

   a. 12

   b. 14

   c. 16

   d. 18

7. Except for special types of stairways, such as spiral, a required headroom of ______ must be maintained in all parts of the stairway.

   a. 6 feet 6 inches

   b. 6 feet 8 inches

   c. 7 feet 0 inches

   d. 7 feet 6 inches

8. A smoke-developed index of no more than ______ should be maintained for all wall and ceiling finishes.

   a. 200

   b. 250

   c. 375

   d. 450

Study Tip

Smoke-developed rating is a numerical index that indicates the relative density of smoke produced by burning assigned to a specific material.

Study Tip

Habitable space is one with the intended purpose of living, sleeping, eating or cooking. Bathrooms, closets etc., would not be considered such.

9. Except for kitchens, habitable rooms shall not be less than ______ feet in any horizontal dimension.

   a. 6
   b. 7
   c. 8
   d. 9

10. A 10' × 10' habitable room should have an aggregate glazed area of ______ square feet to provide natural lighting.

    a. 5
    b. 6
    c. 7
    d. 8

11. Not more than ______ of the required floor area of a room included as square feet is allowed to have a sloped ceiling height of 7'.

    a. 30%
    b. 40%
    c. 50%
    d. 60%

12. Foam plastic picture mold, chair rails, baseboards, handrails, ceiling beams, and similar trim are allowed by code provided the minimum density is:

    a. 15 pcf
    b. 20 pcf
    c. 25 pcf
    d. 30 pcf

13. All rescue openings are required to have at least a net clear opening of ______ square feet unless situated on a grade level.

    a. 5.7
    b. 5.0
    c. 4.7
    d. None of the above

14. The maximum flame spread is ______ for any structure's wall and ceiling finishes. This does not apply to most trim finishes.

    a. 75
    b. 150
    c. 200
    d. 250

15. In the event an emergency escape and rescue opening is positioned with a sill height below the adjacent grade, a window well is required. The minimum horizontal area of the well should be ______ square feet.

    a. 8
    b. 9
    c. 10
    d. 12

16. For the purpose of the code, a vehicle covering is considered a carport if it is open:

    a. On at least one side other than the entry
    b. On all four sides
    c. On at least two sides
    d. On at least one side and not adjacent to the structure

Study Tip

MINIMUM *opening dimensions will not multiply to produce* ***MINIMUM*** *net openings. One of the two* ***MINIMUM*** *dimensions will need to be increased.*

Study Tip

There are two types of gypsum wallboard, regular and type X. Type X wallboard is formulated by adding noncombustible fibers to the gypsum to provide greater resistance to heat transfer during fire exposure.

17. The minimum width of an exit passageway (hallway) for a single residence shall be at least:

    a. 3 feet

    b. 3 feet 6 inches

    c. 4 feet

    d. 4 feet 2 inches

18. A guardrail system is required for protection when a porch or balcony is located more than ______ inches above the grade or floor below.

    a. 24

    b. 30

    c. 36

    d. 40

19. What is the required width of a landing serving a 42"-wide stairway?

    a. 36"

    b. 42"

    c. 48"

    d. Any of the above

20. What size and type gypsum board is required to separate a garage from a habitable room located above it?

    a. ½" Type X

    b. ⅝" Type X

    c. ¾" Type X

    d. $\frac{3}{16}$" Type X

21. Additional safety measures, such as the use of tempered glass, are required when glazing adjacent to stairways is within ______ inches horizontally of the bottom tread of a stairway in any direction and the exposed surface of the glass is less than 5' above the nose of the tread.

    a. 36

    b. 42

    c. 48

    d. 60

22. Handrail height for stairs, when required, should be a minimum of ______ inches and a maximum of ______ inches measured vertically from the sloped plane adjoining the tread nosing.

    a. 38/34 (respectively)

    b. 34/36 (respectively)

    c. 34/38 (respectively)

    d. 36/38 (respectively)

23. Where entry is made to a crawl space for service of utilities, foam plastic insulation shall be protected against ignition by ______ inch-thick corrosion-resistant steel.

    a. 0.024

    b. 0.016

    c. 0.25

    d. 0.033

24. Wood siding installed on the exterior of a structure should have a clearance of ______ inches from the ground.

    a. 4

    b. 6

    c. 8

    d. 10

Study Tip

Tempered glass *is a type of glass that has increased strength and will usually shatter into small fragments when broken.*

Study Tip

A Parapet wall is the term used to refer to the vertical portion of the wall as it extends above a low-slope (flat) roof.

25. The minimum height of a parapet wall where the roof surface is adjacent to the wall or walls at the same elevation is:

    a. 24"

    b. 28"

    c. 30"

    d. 36"

PART FOUR

Foundations

Questions are based on the Building Code.

The following code quiz is based on topics and issues related to foundations. Exam candidates preparing for an exam based on the *International Residential Code* will find the answers in Chapter 4. Candidates using the *International Building Code* will find the answers in various chapters.

Questions in this quiz cover the following subject matter:

- Footings
- Foundation walls
- Drainage
- Waterproofing

Study Tip

Class of Soil is described in a Table at the beginning of the chapter.

1. A plain masonry hollow wall that is 7' tall with 6' of unbalanced backfill should be at least ______ thick. The soil type is GW.

   a. 6"

   b. 8"

   c. 10"

   d. 12"

2. The allowable foundation (load-bearing pressure) in pounds per square foot for sedimentary and foliated backfill is:

   a. 2,000

   b. 3,000

   c. 4,000

   d. 12,000

3. Where masonry veneer is used, concrete foundation walls are required to extend ______ inches above the finished grade adjacent to the foundation walls at all points.

   a. 2

   b. 4

   c. 6

   d. 8

4. A carport slab is to be installed in an area that is subjected to severe weather conditions. The minimum compressive strength of the concrete to be used is ______ psi.

   a. 2,000

   b. 2,500

   c. 3,000

   d. 3,500

5. All exterior footings should be placed at least ______ inches below the undisturbed ground surface.

   a. 6

   b. 8

   c. 10

   d. 12

6. Foundations plates or sills should be attached to concrete foundations with bolts:

   a. ½ inch in diameter with a penetration of at least 7 inches

   b. ¾ inch in diameter with a penetration of at least 7 inches

   c. ½ inch in diameter with a penetration of at least 6 inches

   d. ¾ inch in diameter with a penetration of at least 6 inches

7. The grade adjacent to a structure should be sloped away from the building to create a positive drain. The slope should not be less than:

   a. 5%

   b. 6%

   c. 8%

   d. 10%

8. Openings for under-floor (crawl space) ventilation should be covered provided that the least dimension of the covering does not exceed ______ inch(es).

   a. ¼

   b. ½

   c. ¾

   d. 1

Study Tip

When a slope is described as a percentage, an 8% slope indicates the surface will drop 8 feet in a 100 foot run. For shorter distances, convert to inches. EXAMPLE: a 5% slope for a distance of 10 feet: convert 10 feet to inches ($10 \times 12 = 120$). Take 5% of 120. The slope will drop 6 inches in 10 feet.

Notes

9. Gravel or crushed stone drains installed for the purpose of foundation drainage shall extend at least ______ beyond the outside edge of the footing.

   a. 0.5 ft.

   b. 1 ft.

   c. 1.5 ft.

   d. 2 ft.

10. Concrete in footings should have a specified pounds per square inch of at least:

    a. 1,500

    b. 2,000

    c. 2,500

    d. 3,000

11. Stepped footings are required when the grade falls to cause the bottom surface of the footings to slope more than ______ unit(s) vertical and ten units horizontal.

    a. 0.5

    b. 1

    c. 1.5

    d. 2

12. All permanent supports are not required to be protected from frost. Which of the following is required to have frost protection? NOTE: Not light-framed construction

    a. A freestanding building more than 400 square feet

    b. A freestanding building 400 square feet or less with an eave 10 feet in height

    c. A freestanding building 400 square feet or less with an eave 8 feet in height

    d. A freestanding building 300 square feet with an eave 8 feet in height

13. A wall 10" thick is being supported on a foundation. What is the minimum foundation required to support this wall?

    a. 8"

    b. 10"

    c. 12"

    d. 16"

14. Pier and curtain wall foundations are permitted to support light frame construction not more than two stories. All of the following requirements must be met by code except:

    a. Load-bearing wall shall be placed on continuous concrete footings placed integrally with the exterior wall

    b. Thickness of the load-bearing wall shall not be less than 3⅝" nominal

    c. Maximum height of a 4" load-bearing masonry foundation wall supporting wood framed walls shall be more than 4' in height

    d. Unbalanced fill for 4" foundation walls shall not exceed 24" for solid masonry.

15. A rubble stone foundation wall shall have a minimum thickness of _______ inches.

    a. 8

    b. 10

    c. 12

    d. 16

Study Tip

A Curtain wall *is a term used to describe a building façade which does not carry any weight from the building other than its own dead load.*

Notes

PART FIVE

Floors

Questions are based on the Building Code.

The following code quiz is based on topics and issues related to floor construction. Exam candidates preparing for an exam based on the *International Residential Code* will find the answers in Chapter 5. Candidates using the *International Building Code* will find the answers in various chapters.

Questions in this quiz cover the following subject matter:

- Joist bearing
- Joist sizing
- Boring and notching
- Spans
- Fastening
- Drainage
- Waterproofing

Study Tip

SPF—is Spruce, Pine, and Fir.

YP—stands for Yellow Pine.

1. Joist framing from opposite sides of a beam or girder shall be lapped at least ______ inches.

   a. 2

   b. 3

   c. 4

   d. 6

2. A single header is allowed in a framed floor opening if the dimension along the header is less than ______ feet.

   a. 3

   b. 4

   c. 5

   d. 6

3. SPF #2 is to be installed 16" on center in a residential living area with a dead load of 20 psf. The length of the room is 14' 6". What size floor joist is required?

   a. 2 × 6

   b. 2 × 8

   c. 2 × 10

   d. 2 × 12

4. Holes bored in a floor joist to install plumbing pipes should not be placed closer than ______ from the top.

   a. Two times the thickness of the pipe

   b. 2 inches

   c. One half the joist depth

   d. None of the above

5. In residential structures, when there is usable space above and below the concealed space of a floor/ceiling assembly, draftstops must be installed so that the area of the concealed space does not exceed ______ square feet.

   a. 500

   b. 800

   c. 1,000

   d. 1,200

6. Unless bearing on a beam or similar support, approved hangers are required for framed openings in the floor when the header joist to trimmer joist connections and header joist span exceeds ______ feet.

   a. 4

   b. 5

   c. 6

   d. 12

7. The minimum thickness of a concrete slab placed on grade is ______ inches.

   a. 3

   b. 3.5

   c. 4

   d. 6

8. Sheathing, with a span rating of 24/0, to be used for subfloor can be installed with a maximum span of ______ inches.

   a. 0

   b. 16

   c. 24

   d. 32

Study Tip

Draft Stops are installed to restrict the movement of air within open spaces of building components.

Study Tip

Dead Load refers to the weight of the components that make up a building including the walls, floors, windows and so forth.

9. What is the maximum span for a 2 × 8 Southern yellow pine #2 installed in the sleeping area of a structure? The dead load is 20 psf, and the joist spacing is 24 inches on center.

   a. 12 feet 4 inches

   b. 14 feet 2 inches

   c. 13 feet 6 inches

   d. 11 feet 0 inches

10. Beams and girders must bear at least ______ inch(es) on a masonry column.

    a. 1

    b. 1½

    c. 1¾

    d. 3

11. Joists framing into the side of a wood girder must be supported by:

    a. Framing anchors

    b. 2 × 2 Ledger strips

    c. All of the above

    d. None of the above

12. A notch made at the end of a floor joist should not exceed ______ the depth of the joist.

    a. ⅓

    b. ¼

    c. ⅙

    d. ½

13. A one-story building with a width of 28' requires _______ girder(s). *Note*: The span is 10' 2".

    a. Four 2 × 12

    b. Four 2 × 10

    c. Four 2 × 8

    d. Three 2 × 10

14. The required joint laps for approved vapor barrier to be placed between a concrete floor slab and the grade is _______ inches.

    a. 4

    b. 5

    c. 6

    d. 8

15. The minimum thickness of particle board underlayment shall be _______ inch(es).

    a. ¼

    b. ½

    c. ¾

    d. 1

16. What is the maximum span for a 2 × 8 nominal floor joist (Southern pine #2) to be installed in a sleeping area? *Note*: Dead load is 10 psf, and joists will be installed 16 inches on center.

    a. 10 feet 9 inches

    b. 14 feet 2 inches

    c. 13 feet 6 inches

    d. 13 feet 3 inches

Study Tip

Live Load refers to the weight of the people and personal items such as furniture that will occupy the building.

Study Tip

Code requirements with regard to boring holes and notching are based on actual dimensions. For example, the allowable percentage should be taken from 11¼" rather 12" on a 2 × 12 nominal size piece of lumber.

17. What is the maximum notch depth allowed for a 2 × 12 nominal floor joist? *Note*: The joist spans 14 feet 6 inches, and the notch needs to be made 5 feet from one end.

    a. 1⅞ inches

    b. 1¾ inches

    c. 2 inches

    d. None of the above

18. Southern yellow pine #2 is to be installed in a residential sleeping area 16" on center. The span will be 18', and the dead load is 10 psf. What size floor joist is required?

    a. 2 × 6

    b. 2 × 8

    c. 2 × 10

    d. 2 × 12

19. What type lumber is required for a sleeping area with a 20 psf dead load if a 2 × 10 is to span 18' with 16" on center spacing? *Note*: The live load is 30 psf.

    a. Douglas fir-larch ss

    b. Hem-fir ss

    c. Southern pine ss

    d. None of the above

20. A hole to be bored into a floor joist for the purpose of running a plumbing pipe or electrical conduit:

    a. Must not exceed one third the depth of the joist

    b. Must not exceed one sixth the depth of the joist

    c. Cannot be bored in the middle third of the joist

    d. Is not allowed unless approved by a qualified engineer

21. How many jack studs are required to support three 2 × 12s that span 10' 2" on a 20' wide building?

    a. One on each end

    b. Two on each end

    c. Three on each end

    d. Four on each end

22. When joists exceed a nominal 2 × 12" (6:1 depth-to-thickness ratio), bridging must be installed for each ______ feet of span.

    a. 4

    b. 8

    c. 12

    d. 16

23. Without approval from a registered design professional, which of the following jobsite modifications can be made to wood floor trusses?

    a. Cut

    b. Spliced

    c. Notched

    d. None of the above

24. Vapor retarder is not required between the base course / subgrade and a concrete floor slab in which of the following circumstances?

    a. Typical driveway and walkways

    b. Unheated accessory structures

    c. For installations approved by the building official based on local site conditions

    d. All of the above

Study Tip

Vapor Retarders limit the moisture vapor that pass through a wall or floor assembly.

Study Tip

PSF indicates the Pounds per Square Foot.

25. The specified residential sleeping area live load is ______ psf.

a. 10

b. 20

c. 30

d. 40

PART SIX

Walls

Questions are based on the Building Code.

The following code quiz is based on topics and issues related to wall construction. Exam candidates preparing for an exam based on the *International Residential Code* will find the answers in Chapter 6. Candidates using the *International Building Code* will find the answers in various chapters.

Questions in this quiz cover the following subject matter:

- Stud spacing
- Cutting and notching
- Bearing and support
- Fastening

Study Tip

When using Tables in the Code Book, be sure to identify each important element in the question. Remember to be very careful and methodical.

1. Where rafters are spaced more than 16 inches on center and the bearing studs below are spaced 24 inches on center, members shall bear within ______ inches of the studs beneath.

   a. 4

   b. 5

   c. 6

   d. 7

2. What is the maximum spacing of a 2 × 6 for a bearing wall to support a roof and ceiling only?

   a. 16 inches on center

   b. 20 inches on center

   c. 24 inches on center

   d. 28 inches on center

3. Where horizontal steel strap blocking is used to provide edge blocking, it shall be at least ______ inch(es) wide.

   a. 1

   b. 1¼

   c. 1½

   d. 2

4. What is the maximum stud spacing if particleboard wall sheathing is to be nailed to the stud walls? *Note*: M-2 Exterior glue grade, ½ inch thickness.

   a. 12 inches on center

   b. 16 inches on center

   c. 24 inches on center

   d. 30 inches on center

5. Fireblocking is required horizontally at intervals not exceeding ______ feet to cut off all concealed draft openings in framed walls in order to form an effective fire barrier between stories.

   a. 5

   b. 8

   c. 10

   d. 12

6. Foundation cripple walls are required to be framed of studs not less in size than the studding above with a minimum length of ______ inches.

   a. 10

   b. 12

   c. 14

   d. 16

7. When notching of a top plate is required for the purpose of placing piping or ductwork, a galvanized metal tie must be fastened to each plate across and to each side of the opening. The metal tie must be fastened with ______ nails and must be ______ wide.

   a. 12d/½ inch

   b. 12d/1½ inch

   c. 16d/½ inch

   d. 16d/1½ inch

8. A hole may not be bored closer than ______ inch to the edge of a stud located in an exterior load-bearing framed wall.

   a. ¼

   b. ½

   c. ¾

   d. ⅝

Study Tip

Fireblocking can be referred to as Draft Stops or simply Blocking on an exam or in the Code Book.

Study Tip

Cold-formed steel is formed at room temperature as opposed to hot-rolled steel being formed at elevated temperatures. Cold-formed steel is much thinner, therefore, buckling must be considered.

9. The minimum thickness of cold-formed steel studs is ______ mils.

a. 0.024

b. 0.033

c. 0.043

d. 0.054

10. Fasteners for attaching structural sheathing (shear panels) to steel wall framing must be installed with a minimum edge distance of ______ inch.

a. ¼

b. ½

c. ⅜

d. 1

11. The minimum flange width for a load-bearing cold-formed steel stud is ______ inches.

a. 1½

b. 1⅝

c. 1¾

d. 2

12. Special instructions apply for splicing and cripple wall installations where stepped footings cause the height of a required braced wall panel extending from the foundation to the floor above to vary more than ______ feet.

a. 4

b. 6

c. 8

d. 10

13. Buildings must be provided with exterior and interior braced wall lines. The spacing of the braced wall lines shall not exceed ______ feet in both the longitudinal and transverse directions in each story. *Note*: No exceptions apply.

    a. 10

    b. 25

    c. 35

    d. 50

14. The maximum-size hole that can be placed in any stud is ______ inches.

    a. 1.4

    b. 1.6

    c. 2

    d. 2.2

15. What is the maximum spacing of 2 × 4's to be installed in a bearing wall to support one floor, roof, and ceiling?

    a. 10"

    b. 14"

    c. 16"

    d. 24"

16. Particleboard (M-1/M-S Grade) wall sheathing nailed to 16 inches on center studs must be at least ______ inch(es) in thickness.

    a. ½

    b. ⅜

    c. ⅝

    d. 1

Study Tip

MS grade Particle Board was developed by the particleboard industry in response to the need for a lower cost, lighter weight panel with slightly reduced strength properties.

Study Tip

A cripple wall is one that is less than full story height and usually occurs between the first floor and the foundations.

17. A 4-foot opening to be located in an interior nonbearing partition may be constructed with a ____________ header. *Note*: Vertical distance above the opening to the parallel nailing surface above is 24 inches.

    a. Single stud on each side to support a flat 2 × 4
    b. Double stud, single header
    c. Double stud supporting two flat 2 × 4's
    d. None of the above

18. Interior partitions (nonbearing) may be constructed with at least a:

    a. Double top plate
    b. Single top plate
    c. Double top plate if wall height exceeds 10 feet
    d. None of the above

19. Cripple walls having a stud height exceeding ______ inches shall be braced or shall be framed of solid blocking.

    a. 14
    b. 24
    c. 36
    d. 48

20. When single top plates are permitted, the plate must be adequately tied at joints, corners, and intersecting walls by a galvanized steel plate that is nailed to each wall or segment by ______ nails on each side.

    a. Eight 16d
    b. Four 16d
    c. Six 8d
    d. Four 8d

21. The minimum-size screw to attach steel sheets to metal framing (steel to steel) is:

    a. #5
    b. #6
    c. #7
    d. #8

22. What is required when nailing a top plate to a stud?

    a. Two 16d common
    b. Three 16d common
    c. Two 8d common
    d. Three 8d common

23. Which of the following materials is *not* allowed for fireblocking required in wall framing?

    a. 2" nominal lumber
    b. ½" particleboard with joints backed by ½" particleboard
    c. Properly installed ½" gypsum board
    d. Two thicknesses of 1" nominal lumber with broken lap joints

24. Any stud in an exterior wall or bearing partition may be cut or notched to a depth not to exceed ______ of its width. *Note*: No exceptions.

    a. 15%
    b. 25%
    c. 40%
    d. 60%

Study Tip

Nail sizes are believed to be based on a term used in the early 1600's in England. 100 small nails that sold for 4 pence were called 4d nails. (4 d is the abbreviation of 4 pence)

Study Tip

Utility grade studs are relatively low-strength members and often look pretty rough.

25. Utility-grade studs cannot be spaced more than ______ inches on center.

    a. 12

    b. 16

    c. 20

    d. 24

PART SEVEN

Wall Coverings

Questions are based on the Building Code.

The following code quiz is based on topics and issues related to wall covering. There are two quizzes in this section, one based on the IRC and one based on the IBC. Exam candidates preparing for an exam based on the *International Residential Code* will find the answers in Chapter 7. Candidates using the *International Building Code* will find the answers in Chapter 14.

Questions in this quiz cover the following subject matter:

- Gypsum
- Masonry veneer
- Aluminum siding
- Plaster

Study Tip

The first 10 questions in this section are addressed in the International Residential Code. If you are using the International Building Code, proceed to page 87.

The following quiz is based on the ***International Residential Code.***

1. What is the maximum support spacing for ½" gypsum on walls or ceilings?

   a. 16"

   b. 24"

   c. 30"

   d. 36"

2. Shakes used for exterior covering should be held in place by which of the following type fasteners?

   a. Stainless steel

   b. Aluminum nails

   c. Staples

   d. All of the above

3. What is the nominal thickness of horizontal aluminum siding with insulation?

   a. 0.019 inch

   b. 0.024 inch

   c. 7⁄16 inch

   d. 2 inches

4. A weep screed should be provided at or below the foundation plate line on exterior stud walls and shall be placed a minimum of ______ inch(es) above paved areas.

   a. 1

   b. 2

   c. 3

   d. 4

5. What is the minimum number of plaster coats required when applied over metal lathe?

    a. 1
    b. 2
    c. 3
    d. 4

6. Approved corrosive-resistive flashing should be provided in the exterior wall envelope at all the following locations *except*:

    a. At wall and roof intersections
    b. At built-in gutters
    c. Over metal copings and sills
    d. Continuously above all projecting wood trim

7. Gypsum board utilized as the base or backer for adhesive application of ceramic tile should conform to:

    a. ASTM C 1178
    b. ASTM C 94
    c. ASTM C 1395
    d. ASTM C 1047

8. What is the maximum spacing of screws allowed for ⅜" gypsum board installed on a wall with adhesive?

    a. 7"
    b. 12"
    c. 16"
    d. 24"

Study Tip

ASTM, American Society for Testing and Materials, is a standards developing organization that develops and publishes voluntary technical standards.

Study Tip

Water-Resistant gypsum board may be referred to as MR on plans and in test questions. MR stands for Moisture Resistant.

9. Water-resistant gypsum board should *not* be used in which of the following locations?

   a. Shower

   b. Bathtub

   c. Area subject to high humidity

   d. All of the above

10. Masonry veneer lintels should *not* support any vertical load other than:

    a. Live loads

    b. Dead loads

    c. Nonloadbearing

    d. Loadbearing

The following quiz is based on the ***International Building Code.***

Study Tip

The next ten questions are based on the International Building Code. If you are using the International Residential Code, proceed to the next chapter.

1. Stone veneer units not exceeding 10" in thickness should be anchored directly to each of the following *except*:

   a. Masonry

   b. Concrete

   c. Stud construction

   d. All of the above

2. Exterior walls should provide weather protection for the building. Which of the following types of covering is required to have a minimum thickness of ¼"?

   a. Anchored masonry veneer

   b. Vinyl siding

   c. Adhered terra cotta

   d. Precast stone facing

3. The height of any section of thin exterior structural glass veneer should not exceed _______ inches.

   a. 12

   b. 18

   c. 36

   d. 48

4. What is the maximum weight allowed by code of interior adhered masonry veneers?

   a. 15 psf

   b. 20 psf

   c. 25 psf

   d. 30 psf

Notes

5. A mixture of calcined gypsum, or calcined gypsum, lime, and aggregate is defined as:

   a. Gypsum plaster

   b. Cement plaster

   c. Gypsum veneer plaster

   d. None of the above

6. Wood veneers on exterior walls should not project more than ______ inches from the building wall.

   a. 12

   b. 18

   c. 24

   d. 30

7. Nails used to fasten vinyl siding and accessories should have a minimum shank diameter of ______ inch(es).

   a. 0.313

   b. 0.125

   c. 0.75

   d. 1

8. Vinyl siding installed vertically should have fastener spacing not to exceed ______ inches.

   a. 12

   b. 16

   c. 18

   d. 24

9. Flashing should be installed in which of the following locations?

   a. Chimneys

   b. Decks

   c. Built-in gutters

   d. All of the above

10. Fiber cement horizontal lap siding should be lapped a minimum of _______ inch(es) and located over a strip of flashing.

    a. ½

    b. ¾

    c. 1

    d. 1¼

Study Tip

Flashing is a building component that is stalled to prevent water seepage.

Notes

PART EIGHT

Roof Framing

Questions are based on the Building Code.

The following code quiz is based on topics and issues related to roof framing construction. Exam candidates preparing for an exam based on the *International Residential Code* will find the answers in Chapter 8. Candidates using the *International Building Code* will find the answers in Chapter 23.

Questions in this quiz cover the following subject matter:

- Rafter and joist spacing
- Fastening
- Cutting and notching
- Ventilation
- Access

Study Tip

Bridging is a method of placing support between joists to distribute loads and add structural stability.

1. For roof framing, rafters, and ceiling joists, when the nominal depth-to-thickness ratio of the framing member exceeds 6:1, bridging must be installed every ______ feet.

   a. 6

   b. 7

   c. 8

   d. 10

2. A 2 × 6 SYP #2 material is to be used for the ceiling joists with a live load of 10 psf. The joists, spaced 24 inches on center, will be installed to support an attic that will not be used for storage. What is the maximum span?

   a. 11 feet 0 inches

   b. 15 feet 6 inches

   c. 15 feet 11 inches

   d. 17 feet 8 inches

3. Ceiling joists shall have a bearing surface of not less than ______ inch(es) on the wood top plate.

   a. 1

   b. 1½

   c. 2

   d. 2½

4. For ceilings not attached to rafters with a roof live load of 20 psf and a dead load of 10 psf, the maximum span allowed for a 2 × 8 rafter to be spaced 16 inches on center is ______. *Note:* Material is SYP #2.

   a. 18 feet 6 inches

   b. 19 feet 5 inches

   c. 14 feet 4 inches

   d. 19 feet 3 inches

5. For ceilings attached to rafters with a roof live load of 20 psf and a dead load of 20 psf, the maximum span allowed for a 2 × 8 rafter to be spaced 12 inches on center is _______. *Note:* Material is Douglas Fir #1.

   a. 18 feet 6 inches

   b. 19 feet 5 inches

   c. 14 feet 4 inches

   d. 19 feet 3 inches

6. A notch must be placed in a ceiling joist. The nominal size of the joist is 2 × 10 and spans 12 feet 8 inches. What is the maximum notch that can be placed on the top, 2 feet from the exterior wall?

   a. 1.67 inches

   b. 1.54 inches

   c. 1.33 inches

   d. 0 inches

7. How many 16d common nails are required for a rafter connection if the roof slope is 4:12? Rafters are 16" on center, the roof span is 20', and the ground snow load is 30 psf.

   a. 4

   b. 6

   c. 8

   d. 10

8. What type of nails can be substituted for 16d nails when fastening rafter connections?

   a. 10d

   b. 12d

   c. 40d box

   d. No substitutions provided by code

Study Tip

When you have multiple variables in a question, such as number 7, you will be looking for a chart or table to answer the question.

Study Tip

A purlin *is a horizontal structural member that supports roof loads and transfers them to roof beams.*

9. A 24' rafter spaced 24" on center must have a connection strength for resisting uplift winds of ______ pounds considering a basic wind speed of 90 mph.

   a. –145
   b. –181
   c. –262
   d. –351

10. When a ceiling joist must be spliced, what is the least amount it must be lapped?

   a. 2"
   b. 3"
   c. 4"
   d. 6"

11. Purlins used to support roof loads must be at least:

   a. 2 × 6
   b. One size larger than the rafter it is bracing
   c. No less than one size smaller than the rafter it is bracing
   d. The size of the rafter it is supporting

12. For buildings with a roof pitch of less than ______, hips and valleys must be designed as beams.

   a. 2:12
   b. 3:12
   c. 4:12
   d. 5:12

13. A fireplace opening that requires a header greater than ______ must be supported by approved rafter hangers.

    a. 4'

    b. 5'

    c. 6'

    d. 7'

14. What is the minimum net thickness of roof sheathing (surfaced dry) if the framing is 24" on center?

    a. ½"

    b. ⅜"

    c. ⅝"

    d. 7⁄16"

15. A hole must be bored in a ceiling joist and needs to be placed near the edge. What is the closest the hole can be placed to the edge?

    a. 1"

    b. 2"

    c. 2½"

    d. 3"

16. The unbraced length of braces supporting purlins shall not exceed ______.

    a. 4'

    b. 6'

    c. 8'

    d. 10'

Study Tip

A brace is a member used to stiffen or support a structure.

Study Tip

Fire-Retardant-Wood is wood that has been treated to control the spread of flame, smoke and devastation caused by fire.

17. What species of lumber must be used if 2 × 6 nominal ceiling joists are to span 14' 8" and be spaced 16" on center? The live load is 20 psf, and the application is uninhabitable attics with limited storage.

    a. Douglas fir-larch

    b. Hem-fir

    c. Southern pine

    d. Spruce-pine-fir

18. What is the maximum span for 2 × 8 ceiling joists, 16 inches on center, to support an uninhabited attic with limited storage? *Note:* Dead load is 10 psf, and roof live load is 20 psf.

    a. 23 feet 6 inches

    b. 21 feet 7 inches

    c. 20 feet 10 inches

    d. 19 feet 7 inches

19. What is the minimum required headroom (unobstructed) in the attic space above the access opening?

    a. 30"

    b. 36"

    c. 48"

    d. 72"

20. Each of the following must be listed on the label for fire-retardant-wood *except:*

    a. ID mark of the approving agency and treating manufacturer

    b. Name of the fire-retardant treatment and species of wood

    c. Flame and smoke spread index

    d. Statement of compliance with ASTM E 184

21. The net free ventilation area for attics must be at least ______ of the area. *Note:* No exceptions.

    a. 1/150

    b. 1/200

    c. 1/250

    d. 1/300

22. The maximum-allowed moisture content for lumber used to frame a 4:12 pitch roof is:

    a. 15%

    b. 19%

    c. 22%

    d. 27%

23. A listed flame spread index of ______ or less is required when codes specify fire-retardant-treated wood.

    a. 15

    b. 20

    c. 25

    d. 30

24. To prevent ventilation obstruction, ______ inch(es) of space must be maintained between the insulation and sheathing where eave or cornice vents are installed.

    a. 1

    b. 1½

    c. 2

    d. 2¾

Study Tip

Flame Spread Index is the numeric value assigned to a material to indicate the degree to which the fire will propagate over a surface.

Study Tip

When referring to a table in the Code Book, be sure to carefully identify each section and compare it to each component of the question.

25. What is the maximum span of a 2 × 6 SPF #2 ceiling joist to be spaced 12 inches on center? The attic will not be utilized in any way. Dead load is 5 psf, and the live load is 10 psf.

    a. 18 feet 8 inches

    b. 14 feet 9 inches

    c. 16 feet 3 inches

    d. 13 feet 9 inches

PART NINE

Roof Coverings

Questions are based on the Building Code.

The following code quiz is based on topics and issues related to roofing assemblies. Exam candidates preparing for an exam based on the *International Residential Code* will find the answers in Chapter 9. Candidates using the *International Building Code* will find the answers in Chapter 15.

Questions in this quiz cover the following subject matter:

- Asphalt shingles
- Wood shakes
- Mineral-surfaced roll roofing
- Underlayment
- Attachment
- Valley
- Flashing

Study Tip

The Roof Covering section of the Code Book is divided into sections that address each type of covering. Identify the covering type, such as asphalt, and then scan looking for the answer to the question such as slope or fastening requirement.

1. What is the minimum slope of asphalt shingles that may be used on a roof?

   a. 2:12

   b. 3:12

   c. 4:12

   d. 6:12

2. All of the following fasteners are used for asphalt shingles *except:*

   a. Galvanized steel

   b. Copper

   c. Aluminum

   d. Nickel

3. The minimum-diameter head permitted to be on a fastener for asphalt shingles is ______ inches.

   a. 0.105

   b. 0.375

   c. 0.50

   d. 0.75

4. Asphalt shingles for normal application shall be secured with not less than ______ fastener(s) per shingle.

   a. 1

   b. 2

   c. 3

   d. 4

5. *Given:* A roof slopes 17%, and underlayment is being installed. The first layer of underlayment is a minimum of 19 inches wide, and the second layer shall be ______ inches wide.

   a. 14

   b. 24

   c. 36

   d. 42

6. Fasteners for underlayment installed in an area with high winds shall be applied along the overlap not farther than ______ inches apart on center.

   a. 12

   b. 24

   c. 36

   d. 42

7. On roofs sloped 4:12 or greater, underlayment shall be at least one-layer applied shingle fashion starting from the eaves and lapped ______ inch(es).

   a. 1

   b. 2

   c. 4

   d. 5

8. Base flashing shall have a minimum thickness of ______ inch.

   a. 0.016

   b. 0.017

   c. 0.018

   d. 0.019

Study Tip

Underlayment *is often referred to in the field as "Felt."*

Study Tip

In an open valley, roof coverings do not extend across the valley. The valley flashing is exposed.

9. For closed valleys, valley lining of one ply of smooth roll roofing must be at least ______ inches wide.

   a. 18
   b. 24
   c. 36
   d. 42

10. Clay tile shall be installed on a roof with a minimum slope of:

    a. 2½:12
    b. 3:12
    c. 4:12
    d. 5:12

11. Attachment wire for clay or concrete tile shall not be smaller than ______ inch.

    a. 0.062
    b. 0.077
    c. 0.083
    d. 0.094

12. Which of the following is a consideration when applying tile?

    a. Climate conditions
    b. Roof slope
    c. Type of tile
    d. All of the above

13. When installing metal roof shingles, an ice barrier is required if the average temperature in January is ______ degrees F.

    a. 25

    b. 30

    c. 32

    d. 35

14. When using metal roof shingles, the minimum height of a splash diverter required for roof valley flashing is ______ inch.

    a. ¼

    b. ½

    c. ¾

    d. 1

15. When using spaced sheathing for wood shingles, what are the minimum dimensions required?

    a. 1" × 4"

    b. 1½" × 3"

    c. 1" × 5"

    d. 1½" × 5"

16. What is the minimum head lap allowed for slate shingles secured to a roof with 6:12 slope?

    a. 1"

    b. 2"

    c. 3"

    d. 4"

Study Tip

In a closed valley, the roof covering is lapped from both sides of the valley. The valley is covered or "closed."

Study Tip

Hand-split shakes either come in Premium Grade or Grade 1. Taper-sawn shakes come in three grades: Premium, Grade 1, and Grade 2.

17. Wood shakes shall be laid with a minimum side lap not less than ______ inch(es).

    a. ¾
    b. 1
    c. 1¼
    d. 1½

18. Which of the following types of wood shakes are not approved for grade No. 1?

    a. 18" shakes of naturally durable wood with 7.5" exposure
    b. 24" taper-sawn shakes of naturally durable wood with 7.5" exposure
    c. 24" taper-sawn shakes of naturally durable wood with 10" exposure
    d. 18" preservative-treated taper shakes of Southern yellow pine with 7.5" exposure

19. Which of the following types of fasteners shall be used for copper roofs?

    a. Galvanized
    b. Copper
    c. Stainless steel
    d. 300 Series stainless steel

20. A metal roof of hard lead shall have a standard thickness of:

    a. 4 psf
    b. 3 psf
    c. 2 psf
    d. 1 psf

21. Mineral-surfaced roll roofing shall not be applied on roof slopes with less than _______ slope.

    a. 2%
    b. 8%
    c. 12%
    d. 33%

22. What is the minimum design slope of a coal-tar built-up roof for drainage?

    a. 8%
    b. 6%
    c. 2%
    d. 1%

23. A sprayed polyurethane foam roof requires a protective coating to be applied no later than _______ after foam application.

    a. 2 hours
    b. 12 hours
    c. 72 hours
    d. Can be applied immediately

24. New roof coverings shall not be installed without first removing all existing layers of roof covering *except:*

    a. Where roof covering is water soaked
    b. Where clay tile is installed over existing wood shakes
    c. Where existing roof has two or more applications of roof covering
    d. Where existing roof covering is asbestos-cement tile

Study Tip

Mineral Surfaced Roll Roofing is a general purpose roofing product that utilizes hot mopping asphalt, cold process adhesives and mechanical fasteners.

Study Tip

Make sure you are looking in the correct reference when you are taking your exam. If the question is about theory, it will not be found in your Code Book.

25. Coal-tar saturated organic felt used for built-up roof cover must conform to what standard?

    a. ASTM D 227

    b. ASTM D 450

    c. ASTM D 43

    d. ASTM D 4022

PART TEN

Chimneys and Fireplaces

Questions are based on the Building Code.

The following code quiz is based on topics and issues related to chimney and fireplace construction. Exam candidates preparing for an exam based on the *International Residential Code* will find the answers in Chapter 10. Candidates using the *International Building Code* will find the answers in Chapter 21.

Questions in this quiz cover the following subject matter:

- Chimney termination
- Flue lining
- Hearth size
- Footings
- Smoke chambers
- Firebox
- Framing

Study Tip

The flue is the chimney passage through which the smoke and gasses rise. Its design is critical to create the proper and most effective draft.

1. Footings for masonry chimneys in areas not subjected to freezing shall be at least _____ inches below finished grade.

   a. 6

   b. 12

   c. 18

   d. 24

2. A chimney wall may be reduced in size 4" above where the chimney passes through the roof if:

   a. Proper measurements are taken to prevent excess air from entering

   b. All combustible materials are more than 12" away from chimney wall

   c. The chimney wall is lined with a fire-retardant material

   d. Not permitted under any circumstance

3. Masonry chimneys constructed of hollow masonry units shall be grouted solid with not less than _____ inches of nominal thickness.

   a. 2

   b. 3

   c. 4

   d. 6

4. Masonry chimneys shall be lined appropriately for the type of appliance connected. Residential-type appliances having a clay flue lining must comply with the requirements of:

   a. ASTM C 27

   b. ASTM C 199

   c. ASTM C 315

   d. ASTM 1261

5. Where a spark arrestor is installed on a masonry chimney, the arrestor screen shall have heat and corrosion resistance equivalent to:

   a. 19-gauge stainless steel

   b. 24-gauge stainless steel

   c. 28-gauge stainless steel

   d. 30-gauge stainless steel

6. Chimney terminations shall be not less than _____ feet above the highest point where the chimney passes through the roof.

   a. 2

   b. 3

   c. 6

   d. 10

7. When installing flue lining, the lining shall be carried up vertically with a maximum slope no greater than _____ degrees from the vertical.

   a. 30

   b. 35

   c. 40

   d. 45

8. A 10" round chimney flue venting multiple appliances shall have a cross-sectional area of ________ square feet.

   a. 28

   b. 38

   c. 50

   d. 78

Study Tip

Spark arrestors *do not eliminate sparks but greatly reduce the hazard.*

Study Tip

Draft *is the upward movement of air within the chimney.*

9. The upper edge of a chimney cleanout shall be located at least ___ inches below the lowest chimney inlet opening.

   a. 2

   b. 4

   c. 6

   d. 8

10. Chimneys that pass through the soffit or cornice shall have a minimum air space clearance of ________ inch(es).

    a. 1

    b. 2

    c. 3

    d. 4

11. *Given:* The floor of combustion chamber to the top of the flue is 18 feet. The fireplace opening is 2,200 square inches. What is the minimum cross-sectional area required for a retangular chimney flue?

    a. 125 square inches

    b. 140 square inches

    c. 168 square inches

    d. 214 square inches

12. Fireblocking shall be to a depth of 1" and shall be placed on strips of metal between chimneys and:

    a. Wood joists

    b. Beams

    c. Headers

    d. All of the above

13. Footings for masonry fireplaces shall extend at least ____ inches beyond the face of the fireplace.

    a. 4
    b. 6
    c. 10
    d. 12

14. Each strap used for seismic anchorage shall be fastened to a minimum of four floor-ceiling or floor joists using:

    a. one 1½" bolt
    b. two 1½" bolts
    c. one ½" bolt
    d. Two ½" bolts

15. What is the minimum depth of a masonry chimney firebox?

    a. 20"
    b. 24"
    c. 36"
    d. 30"

16. A fireplace throat shall be located a minimum of _____ inches above the lintel.

    a. 4
    b. 6
    c. 8
    d. 10

Study Tip

The firebox is the part of the fireplace where fuel is combusted.

Study Tip

The smoke chamber is the portion of the fireplace that is located above the firebox at the base of the chimney flue where smoke gathers before it exits the chimney.

17. Smoke chamber walls shall be constructed of which of the following?

    a. Grouted stone

    b. Grouted concrete

    c. Hollow masonry units

    d. All of the above

18. The inside of a fireplace opening is 4 square feet. The inside height of the smoke chamber from the fireplace throat to the beginning of the flue shall not be greater than _____ square feet.

    a. 3

    b. 4

    c. 5

    d. 6

19. What is the absolute minimum thickness of a hearth extension?

    a. ⅜"

    b. 2"

    c. 4"

    d. 8"

20. A hearth extension shall extend at least 20 inches in front of the fireplace opening if the opening is:

    a. Larger than 2 sq. ft.

    b. Larger that 4 sq. ft.

    c. Less than 6 sq. ft.

    d. At least 6 sq. ft.

21. Exterior air intake shall not be located in any of the following areas *except*:

    a. A garage

    b. Outside a dwelling

    c. A basement

    d. An attic

22. Many wythes must be built between adjacent flue linings and shall be at least ___ inches thick and bonded into the walls of the chimney.

    a. 3

    b. 4

    c. 5

    d. 6

23. What is the cross section of a rectangular flue size with a 7½ inches × 15½ inches outside dimension?

    a. 58 square inches

    b. 74 square inches

    c. 82 square inches

    d. 91 square inches

24. Chimneys located inside a building shall have a minimum airspace combustibles of ______ inches.

    a. 2

    b. 4

    c. 6

    d. 12

Study Tip

A masonry wythe *refers to the vertical stack of bricks. A single wythe wall would have multiple* courses *of brick.*

Study Tip

Corbelling is the process of pulling a course of bricks outward, each slightly a bit more than the one below it, to create a shelf-like effect.

25. A chimney shall not be corbelled more than _____ times the chimney's wall thickness from a wall.

    a. 0.25

    b. 0.375

    c. 0.5

    d. 0.75

PART ELEVEN

Concrete

Questions are based on general knowledge of concrete.

The following quiz is based on topics and issues based on concrete theory. As opposed to questions based on code, these address the why and how. Refer to your candidate information bulletin for the approved reference(s) that relate to concrete. Each of the questions in this quiz is based on general information about concrete, which can be found in many approved reference books. The answers to these questions will *not* be answered in the code book.

Questions in this quiz cover the following subject matter:

- Types of cement
- Joints
- Curing
- Forms
- Reinforcement
- Slump
- Cover

Study Tip

Concrete is made up of four parts which include: cement, sand, aggregate and water. Many times, sand is referred to as "coarse aggregate."

1. Typically, the required cover for reinforcement in concrete footings is _______ inches.

    a. 1½

    b. 2

    c. 2½

    d. 3

2. The space between the end of the rebar and the form is referred to as _______ .

    a. cover

    b. bond

    c. space

    d. void

3. For each inch of reduced slump in concrete, the pounds per square inch is reduced by _______ .

    a. 100

    b. 125

    c. 150

    d. 200

4. The load limit at which steel breaks is ____________________ .

    a. Yield strength

    b. Ultimate tensile strength

    c. Bond strength

    d. None of the above

5. The best type of Portland cement to use when the strength gain of concrete should be rapid is ________ .

   a. Type I
   b. Type II
   c. Type III
   d. Type IV

6. The most important benefit of air-entrained concrete is?

   a. shorter set up time
   b. improves resistance to freeze/thaw cycles
   c. moves through chute more rapidly
   d. less expensive

7. What is the diameter of a #3 deformed piece of steel reinforcement?

   a. 3"
   b. ⅓"
   c. ⅜"
   d. 0.33"

8. When testing the compressive strength of concrete cylinders, at what point is the test performed?

   a. 24 hours
   b. 48 hours
   c. 1 week
   d. 28 days

Study Tip

There are five basic types of cement. Each type can effect the curing time differently.

Study Tip

Rebar is sized in ⅛" increments up to #8.

9. What type of nail is recommended for the installation of form material that will be removed?

   a. 8d Common

   b. 16d Common

   c. Double-headed

   d. Any of the above

10. The most common type of reinforcement used for the installation of footings is _______________ .

    a. Deformed steel

    b. Smooth steel

    c. Rigid steel

    d. #3 Smooth steel

11. What admixture is used to achieve a high slump without the reduced pounds per square inch?

    a. Retarder

    b. Calcium

    c. Superplasticizer

    d. All of the above

12. Normal concrete is designed to reach its intended compressive strength within ___ days.

    a. 7

    b. 14

    c. 21

    d. 28

13. The type of concrete joint that will allow vertical as well as horizontal movement and would likely be used around the perimeter of a concrete column is:

    a. Construction
    b. Isolation
    c. Control
    d. Expansion

14. What is the chemical reaction that takes place when water is added to cement?

    a. hydration
    b. re-hydration
    c. bonding
    d. fusion

15. Which of the following is *not* considered an admixture?

    a. Calcium
    b. Retarder
    c. Aggregate
    d. Superplasticizer

16. The most common welded-wire fabric used for residential applications has wires spaced ________ inches apart in each direction.

    a. 2
    b. 4
    c. 6
    d. 8

Study Tip

Anything added to the concrete mixture other than water, sand, aggregate or cement is considered an admixture.

Study Tip

Water that is unsuitable for drinking should not be used for mixing concrete.

17. What is the largest recommended size of aggregate to be used in high-strength concrete?

    a. ⅜"

    b. ½"

    c. 1"

    d. 1½"

18. When pile foundations are cost-prohibitive due to the allowable bearing capacity being low to great depths, which of the following types of foundations are recommended?

    a. Floating

    b. Raft

    c. Mat

    d. Any of the above

19. In extremely hot weather, what is the maximum allowable time usually specified for the delivery time of ready-mixed concrete?

    a. 30 minutes

    b. 60 minutes

    c. 90 minutes

    d. 120 minutes

20. What type of cement is most commonly used for mixing concrete and is intended as general purpose?

    a. Type I

    b. Type II

    c. Type III

    d. Type IV

21. It is estimated the concrete formwork can account for as much as ___ of the cost for a concrete structure.

    a. 20%
    b. 40%
    c. 60%
    d. 75%

22. A drop chute must be used for concrete that will be dropped more than ___ feet.

    a. 3
    b. 4
    c. 5
    d. 6

23. Longitudinal joints installed to control cracking caused by temperature variations in pavement slabs are called _________ joints.

    a. warping
    b. construction
    c. isolation
    d. expansion

24. To increase the slump of concrete by 1 inch, the rule of thumb is to add ___ gallon(s) of water.

    a. 1
    b. 1¼
    c. 1½
    d. 1¾

Study Tip

A slump test is used to measure consistency of concrete. Concrete is placed in a cone, the cone is then removed causing the concrete to "slump" downward. A measurement is taken from the top of the cone to where the concrete settles—this measurement is the slump.

Study Tip

Reinforced concrete combines the tensile strength of steel and the compressive strength of concrete to create stronger concrete.

25. Which of the following admixtures is added to concrete in hot weather for the purpose of providing more set time for placement?

    a. Calcium

    b. Retarder

    c. Water reducer

    d. Accelerator

PART TWELVE

Masonry

Questions are based on general knowledge of masonry.

The following quiz is based on topics and issues based on masonry theory. As opposed to questions based on code, these address the why and how. Refer to your candidate information bulletin for the approved reference(s) that relate to masonry. The answers to these questions will *not* be answered in the code book.

Questions in this quiz cover the following subject matter:

- Types of mortar
- CMU makeup
- Brick application
- Nominal size Versus Actual size
- Types of joints
- Bonds

Notes

1. The type of mortar best suited for use below grade is:

   a. Type K

   b. Type M

   c. Type N

   d. Type S

2. The type of joint that does not require jointing is:

   a. Flush

   b. Extruded

   c. Rough cut

   d. Weathered

3. The process of remixing mortar after it has started to become stiff and unusable is called:

   a. Reworked

   b. Dehydrated

   c. Rewet

   d. Retempered

4. With regard to masonry units, a "nominal" dimension refers to:

   a. Actual dimension

   b. Standard brick sizes

   c. Nominal brick plus the mortar joint

   d. Actual brick size plus the mortar joint

5. The process of placing a masonry unit beyond the vertical placement of the unit below is called:

   a. Benching

   b. Stepping

   c. Corbelling

   d. None of the above

6. Unused mortar should be discarded _____ hour(s) after it has been mixed.

    a. 1

    b. 1.5

    c. 2

    d. 2.5

7. Which of the following types of brick bond patterns would be considered the weakest?

    a. Flemish

    b. Running

    c. Stack

    d. English

8. A white powderlike deposit that develops on masonry walls that is caused by water-soluble salts is called:

    a. Efflorescence

    b. Lime build-up

    c. Mortar residue

    d. None of the above

9. What percentage of block, by weight, does aggregate make up for concrete masonry units?

    a. 50%

    b. 75%

    c. 80%

    d. 90%

Study Tip

Concrete should never be used after 1 hour and thirty minutes from the time it is mixed. Masonry should never be used after (Question 6) from the time it has been mixed.

Study Tip

Nominal dimensions can be thought of as the "name" size as opposed to the actual dimensions.

10. What is the nominal dimension of a standard concrete masonry unit (block)?

    a. $7\frac{5}{8}$" × $7\frac{5}{8}$" × 16"

    b. $7\frac{5}{8}$" × 8" × $15\frac{5}{8}$"

    c. 8" × $7\frac{5}{8}$" × $15\frac{5}{8}$"

    d. 8" × 8" × 16"

11. Which of the following is the correct name for a block designed with molded or sawed openings for the purpose of absorbing sound?

    a. Insulating block

    b. Sound block

    c. Screen block

    d. Decorative block

12. A ______ masonry wall is constructed with two wythes that can react independently of one another and are separated by a continuous airspace of at least 2 inches (typical).

    a. veneer

    b. hollow

    c. cavity

    d. composite

13. What type of mortar is only suitable for use in nonload-bearing masonry?

    a. Type O

    b. Type N

    c. Type S

    d. Type M

14. Hydrated lime that has been formed into putty before packaging is called:

    a. Mixed

    b. Prehydrated

    c. Slaked

    d. Preserved

15. The recommended minimum time for machine-mixing ingredients for mortar is ______ minutes.

    a. 1 to 3

    b. 3 to 5

    c. 5 to 7

    d. 7 to 9

16. The typical mortar joint size to lay concrete bricks is ____ inch(es).

    a. ½

    b. ⅜

    c. ¾

    d. 1

17. What grade of concrete brick is required when the structure will be exposed to severe frost?

    a. Grade N

    b. Grade S

    c. Grade M

    d. Grade I

Study Tip

If you are taking an "open-book" exam, be sure to identify the section that discusses the different types of mortar and their uses. Try to memorize as little as possible and rely on your references more.

Study Tip

As you are taking the actual exam, remember the questions will be mixed up and not categorized by book. These exercises will help you grab the correct book.

18. Which of the following is used to describe masonry bond?

    a. Mortar joint

    b. Pattern bond

    c. Structural bond

    d. All of the above

19. What type of mortar joint is best when the structure will be subject to heavy rains and high winds?

    a. Weathered

    b. Struck

    c. Flush

    d. V-shaped

20. A brick cut lengthwise across the end is called a _____ closure.

    a. queen

    b. king

    c. bat or half

    d. split

21. Which of the following grades of brick is designed for use as a backup on interior masonry?

    a. Grade NW

    b. Grade AW

    c. Grade MW

    d. Grade SW

22. __________ units are reduced-thickness units, or fire clay, used in veneer applications.

    a. Glazed brick
    b. Thin brick veneer
    c. Green bricks
    d. Paving brick

23. A modular unit is based on a measurement of ___ inch(es).

    a. 1
    b. 3
    c. 4
    d. 5

24. In contrast to a lightweight block, a typical 8" × 16" aggregate and sand block weighs about ___ pounds.

    a. 20
    b. 30
    c. 40
    d. 50

25. Which of the following high-compressive strength mortars is to be used for unreinforced masonry below grade and in contact with the earth?

    a. Type M
    b. Type S
    c. Type N
    d. Type K

Study Tip

Always use your index first by looking up the most specific word in the question.

Notes

PART THIRTEEN

Carpentry

Questions are based on general knowledge of carpentry.

The following quiz is based on topics and issues of basic carpentry. As opposed to questions based on code, these address the why and how. Refer to your candidate information bulletin for the approved reference(s) that relate to masonry. The answers to these questions will *not* be answered in the code book.

Questions in this quiz cover the following subject matter:

- Tools
- Site work
- Framing
- Concrete
- Plans

Study Tip

Plot is a singular reference while Plat is plural. To look at multiple home sites in a subdivision, you would refer to the plat.

1. The location of a residential home as it should be constructed on the building site is specified on the ____ plan.

    a. floor

    b. plot

    c. plat

    d. framing

2. Which of the following floor framing members supports the heaviest load of attached horizontal members?

    a. Joists

    b. Ledger strips

    c. Lintels

    d. Girders

3. What is the standard height for an interior door for residential homes?

    a. 7 feet

    b. 7 feet 2 inches

    c. 6 feet

    d. 6 feet 8 inches

4. The term used to describe the inside of a door opening during the framing stage and prior to the unit installation is:

    a. Net opening

    b. Door opening

    c. Rough opening

    d. Trimmer dimensions

5. What is used to position a leveling instrument directly over a given point during surveying or foundation construction?

   a. Benchmark
   b. Plumb bob
   c. Laser level
   d. Point indicator

6. Which of the following is a common squaring method used to ensure an angle is 90 degrees?

   a. 2-3-6
   b. 2-4-6
   c. 3-4-5
   d. 6-8-9

7. When joists are parallel to a component, such as in a shower or garden tub, the joists should be:

   a. Doubled
   b. Blocked
   c. Braced
   d. Engineered

8. A common field test to verify a rectangular building is properly square at each corner is to:

   a. Make sure the width and length are the same
   b. Check the corners with a building square
   c. Measure the diagonal corners, and if they are equal, the building is square
   d. Check at least one corner using the 3,4,5 system

Study Tip

To ensure an angle is 90 degrees, use the Pythagorean Therom Theory. C = the square root of A squared plus B squared.

Study Tip

Remember to use the index to look up the most specific word in the question. Save yourself time by tabbing the index and using a scratch sheet of paper to write the page number(s) down.

9. What are the three members of a balustrade?
   a. Newels, handrails, and balusters
   b. Treads, risers, and handrails
   c. Stringers, handrails, and newels
   d. Balusters, handrails, and risers

10. What type of saw is used to shape the mitered edges of molding?
    a. Hack
    b. Coping
    c. Jig
    d. Skill

11. What type of saw blade is used for cutting a groove in wood?
    a. Dado
    b. Flat
    c. Cross-cut
    d. Kerf

12. The point at which contractors and surveyors begin an excavation is the ___________ .
    a. level
    b. datum
    c. hub
    d. bench mark

13. Batter boards are typically placed _________________________ .
    a. at the very edge of the foundation lines
    b. 2 feet beyond the edge of the foundation lines
    c. 4 feet beyond the edge of the foundation lines
    d. inside the edge of the foundation lines

14. What size header would be required for a 36" door in a typical wood frame construction application?

    a. 37"

    b. 38½"

    c. 41½"

    d. 43"

15. When pouring a 6" slab, how many square feet will a cubic yard of concrete cover?

    a. 54 square feet

    b. 66 square feet

    c. 76 square feet

    d. 81 square feet

16. To prevent termites from entering a wood sill, use ____________ .

    a. silicone

    b. metal flashing

    c. caulking

    d. termite preventor

17. A ________ is used to produce a rounded edge on a concrete slab.

    a. bull float

    b. edger

    c. screed

    d. spade

Study Tip

To calculate the cubic yards of concrete necessary for a project, Multiply the Length times Width time Depth and divide by 27.

Study Tip

Plywood that is stamped with an acceptable span rating does not mean the rating is acceptable by Building Codes.

18. What is the recommended maximum height of a stepped footing?

    a. 1 foot
    b. 2 feet
    c. 3 feet
    d. 4 feet

19. Plans are to be scaled using ¼ inch = 1 foot. On the plans, how many inches would represent a 34-foot wall?

    a. 4.3 inches
    b. 8.5 inches
    c. 9.6 inches
    d. 12.6 inches

20. If a sheet of plywood is stamped with 36/16, this indicates that the panel is rated for:

    a. Roof framing spaced 16 inches on center and floor framing spaced 36 inches on center
    b. Roof framing spaced 36 inches on center and floor framing spaced 16 inches on center
    c. Roof framing spaced 36 inches on center and floor framing spaced 36 inches on center
    d. Roof framing spaced 16 inches on center and floor framing spaced 16 inches on center

21. What is constructed for the purpose of diverting water around a chimney?

    a. Cricket
    b. Flashing
    c. Valley strips
    d. Diverter

22. Which of the following members is installed to prevent opposing rafters from spreading apart?

    a. Collar tie
    b. Bracing
    c. Rafter jack
    d. Rafter cripples

23. Each of the following statements is *true* about architectural grade glulams *except*:

    a. It is much stronger than a framing grade glulam
    b. All exposed edges are surfaced
    c. Appearance is important but some voids are allowed
    d. Exposed edges are slightly rounded

24. Which of the following statements about cross bridging is *true*?

    a. It is an effective means of bracing
    b. It requires less material
    c. It is not generally required by codes
    d. All of the above

25. The type of drawing to use when precise information is needed about a complex portion of the building is a ____________ .

    a. rendering
    b. schedule
    c. detail
    d. section view

Study Tip

Remember to select the "Best" answer. Many times, a question will have multiple answers that can be true.

Notes

PART FOURTEEN

Roofing

Questions are based on general roofing theory. The first 20 questions can be answered by referring to the Carpentry and Building Construction (Sixth Edition by Feirer) or Roofing Construction & Estimating (by Daniel Atcheson)

The following questions are typical of those found in the trade section of most state contractor licensing exams. Be careful to read your candidate information bulletin to determine whether your exam covers this subject matter and to identify the approved reference material.

Questions in this quiz cover the following subject matter:

- Roofing types
- Roof framing
- Coverings
- Estimating
- Problems

Notes

1. Roof materials are estimated and sold by the square. To convert the net square feet of a roof to squares, divide by:

   a. 10

   b. 50

   c. 100

   d. 1000

2. A roof framing member that will never extend the full distance from the top plate to the ridge board.

   a. Hip jack rafter

   b. Valley jack rafter

   c. Cripple jack rafter

   d. All of the above

3. The board that is nailed to the ends of the rafter tails and is one of the members that makes up the cornice is the:

   a. Fascia

   b. Soffit

   c. Bird's mouth

   d. Frieze board

4. The type of lap in which the shingles or roof covering would be three layers thick is:

   a. Top lap

   b. Head lap

   c. Side lap

   d. End lap

5. The term used to specify the ratio of a roof in terms of its vertical rise to horizontal run.

   a. Pitch
   b. Slope
   c. Factor
   d. All of the above

6. A sheet of roll roofing that becomes the top layer for a Built-up Roof (BUR).

   a. Cap
   b. Top
   c. Drip
   d. None of the above

7. A horizontal member that extends from a rafter end to a nailer or face of the wall sheathing to form a surface to which soffit material is attached.

   a. Cornice
   b. Frieze board
   c. Tail-cut
   d. Lookout

8. The framing member that connects opposite pairs of rafters to help support the roof system is:

   a. Ceiling joists
   b. Collar ties
   c. Cripple jack rafters
   d. Ridge boards

Notes

Notes

9. The portion of the shingle that is not covered by the next course of shingles.

   a. Exposure

   b. Top lap

   c. Side lap

   d. End lap

10. What is the roof factor for determining the roof area from a plan when the vertical rise is 10 inches?

    a. 1.225

    b. 1.275

    c. 1.302

    d. 1.329

11. The result of backup that causes water to stand and create leaks as a result of the freeze/thaw cycles of snow:

    a. Ice shield

    b. Water shield

    c. Ice dam

    d. Water penetration

12. The typical length of a 3-tab asphalt shingle is:

    a. 32"

    b. 36"

    c. 42"

    d. 48"

13. A type of roof valley in which shingles are not applied to the intersection of two roof surfaces.

    a. Open valley

    b. Closed valley

    c. Conventional valley

    d. Flashing

14. The portion of the shingle that is not exposed to the weather but not the shortest distance from the lower edge of the shingle in the second course:

    a. Exposure

    b. Top lap

    c. Head lap

    d. Butt edge

15. Where a chimney (vertical surface) meets the roof, what type of flashing should be used?

    a. Step

    b. Counter

    c. Cap

    d. All three types can be used

16. The type of roof that has a steep slope on two sides in which the second slope begins partway up and continues to the top:

    a. Gambrel

    b. Mansard

    c. Dutch hip

    d. Hip

Notes

Notes

17. Calculate the length of a common rafter to be installed in a Gable Roof system designed with a 6:12 slope. The overhang will be 12" and the span is 24'.

    a. 14'

    b. 15'

    c. 16'

    d. 18'

18. Used to divert water, ice and snow at the upper side of a chimney:

    a. Flashing

    b. Cricket

    c. Saddle

    d. Cricket or saddle

19. What is the Roof-Slope factor for determining the area of a roof if the design calls for a 8 in 12 slope?

    a. 1.054

    b. 1.118

    c. 1.202

    d. 1.302

20. What is the recommended hot-dipped galvanized or aluminum roofing nail head size for fastening asphalt shingles?

    a. ¾"

    b. ⅞"

    c. ⅜"

    d. ⅝"

21. What time of the year is best to repair asphalt shingles?

    a. Summer

    b. Spring

    c. Fall

    d. Winter

22. If the repair costs of a wood shake or shingle roof will exceed ___ percent of its replacement costs, it will be considered "beyond repair."

    a. 50

    b. 60

    c. 70

    d. 80

23. A drip edge should be installed so that it is placed under the underlayment at the eave and ________________________.

    a. the same at the rake

    b. over the underlayment at the rake

    c. either method of installation is recommended

    d. it should be applied over the underlayment at both the eave and the rake

24. The most common type of asphalt:

    a. Type I

    b. Type II

    c. Type III

    d. Type IV

Notes

Notes

25. How many pounds of asphalt bitumen are required per square between felts with an application tolerance of plus or minus 15 percent?

    a. 15

    b. 20

    c. 25

    d. 30

PART FIFTEEN

Project Management

Questions are based on business and project management.

The following questions are typical of those found in the project management section of most state contractor licensing exams. While many state requirements include separate exams to cover this content, several states incorporate these questions intermittently with the trade questions. Be careful to read your candidate information bulletin to determine whether your exam covers this subject matter and to determine the approved reference material.

Questions in this quiz cover the following subject matter:

- Accounting
- Scheduling
- Fair Labor Standards Act (FLSA)
- Business organization
- Payroll
- Federal Unemployment Tax (FUTA)

Study Tip

Pay attention to the advantages and disadvantages of each type of business organization.

Project Management Practice Exam 1

1. Each of the following is regulated by the FLSA *except*:

   a. Vacation

   b. Time and a half for hours over 40

   c. Minimum wage

   d. Compliance with child labor laws

2. What amount does an employer pay for Social Security and Medicare?

   a. 6.2%

   b. 1.45%

   c. 7.65%

   d. 5.4% on the first $7,000

3. Two brothers are planning to start a plumbing contracting company together. The oldest brother will be the primary financial backer and wants no day-to-day involvement in the operations. What type of business should the brothers form?

   a. General partnership

   b. Proprietorship

   c. Limited partnership

   d. Corporation

4. The best source for projecting future project and company overhead percentage rates is:

   a. Historical data

   b. Bid documents

   c. Unit pricing

   d. Job cost records

5. Which of the following insurance types is most widely used for construction projects?

    a. Building operations
    b. Liability
    c. All-risk builder's-risk
    d. Catastrophe

6. Which of the following would be considered project overhead?

    a. Yellow page advertising
    b. Secretary's salary
    c. Electric bill for the office
    d. Electric bill for the project office

7. Yellow Duck Plumbing Company has 21 employees. According to OSHA, this company must keep a record for ______ years after each recorded accident while on the job site for each of its employees.

    a. 3
    b. 4
    c. 5
    d. 6

8. What entity continues to exist beyond the death of one of the principals?

    a. Corporation
    b. Partnership
    c. Sole proprietorship
    d. Limited partnership

Study Tip

Remember an employee NEVER pays FUTA or Worker's Comp. The withholding for either will always be zero.

Study Tip

Typically, an approved reference book for Project Management and Business Law will cover state-specific information in the last half of the book.

9. Employers are not allowed to deduct ______ from employee's wages.

   a. Social Security

   b. Medicare

   c. FUTA

   d. Federal Income Tax

10. Which of the following is a *disadvantage* of setting up a general partnership?

    a. Single taxation

    b. Easy setup

    c. Unlimited personal liability

    d. The inability to pool financial resources

11. Which of the following would *not* constitute a material breach of contract?

    a. Building on the wrong site

    b. A refusal to work

    c. The owner refusing to pay

    d. A delay in the work

12. What form of business organization offers contractors the most protection against personal liability?

    a. Corporation

    b. Partnership

    c. Joint venture

    d. Sole proprietorship

13. *Given:* A carpenter drops a hammer from a roof and injures a curious bystander. Which type of insurance would cover the victim's medical costs?

    a. Builder's risk
    b. Named-peril risk
    c. Liability
    d. Risk management

14. The simplest form of business ownership is ________________.

    a. corporation
    b. partnership
    c. limited partnership
    d. sole proprietorship

15. What form is used to report FUTA?

    a. W-3
    b. 940
    c. W-2
    d. 1099

16. A large plumbing contractor maintains a central office and has crews working at several projects. Which of the following is considered a company overhead cost?

    a. Superintendent salary
    b. Temporary barricades
    c. Site cleanup
    d. Advertising

Study Tip

Direct Costs are those that can be tied to a specific job.

Study Tip

Working capital is the difference between your total annual liabilities and your cash on hand.

17. Typically, the only party with whom subcontractors and their employees have a contractual relationship is the ________.

    a. owner's representative

    b. architect

    c. owner

    d. contractor

18. The purpose of a journal is to:

    a. Post transactions on a weekly basis

    b. Post transactions on a daily basis

    c. Ensure proper job costing

    d. Balance the checkbook

19. What is the latest date an employer can file a FUTA return (Form 940)? *Note*: Assume no time extensions are taken and the balance due is greater than $100.00.

    a. December 1

    b. December 31

    c. January 1

    d. January 31

20. Employee records regarding pay shall be kept for at least ____ year(s).

    a. 1

    b. 2

    c. 3

    d. 4

21. Which of the following is the proper formula for determining assets?

    a. Liabilities + Owner's equity
    b. Liabilities − Owner's equity
    c. Owner's equity − Liabilities
    d. Assets − Liabilities

22. Which of the following components make a change order complete?

    a. The change to be made, the charge, and the effect on time
    b. The owner's/contractor's address and the change order number
    c. The legal description and the contractor's license number
    d. The contractor's license number, the change order number, and the architect's signature

23. The primary purpose of a written contract is to:

    a. Define rights and obligations of each party
    b. Standardize all requirements of work performance
    c. Condense construction requirements
    d. All of the above

24. The term used in scheduling that makes reference to the TIME a project can be delayed without extending the project completion:

    a. Float time
    b. Lag time
    c. Critical task
    d. Noncritical task

Study Tip

A change order is necessary after a contract has been signed. Changes made prior to the contract execution will be made on an addendum.

Study Tip

A contract that is deemed unclear, will always be judged against the party that wrote or provided it.

25. Which of the following terms is used to describe withholding a percentage of a progress payment pending completion to a certain point for the purpose of covering defects or faulty installation?

    a. Retainage

    b. Withholding

    c. Holdback

    d. None of the above

Project Management Practice Exam 2

Notes

1. A bid bond provides:

    a. Assurance that the bonding company is liable for all bid discrepancies

    b. Assurance to the owner or government agency that the contractor will honor the bid submitted by entering into a contract in accordance with the terms of the bid

    c. A guarantee to the owner that the project will be completed

    d. A guarantee of both job completion and payment of all labor and materials

2. A truck that is due to be paid off within a year would be considered a(n):

    a. Current liability

    b. Long-term liability

    c. Asset

    d. Other

3. In the event a subcontractor does not carry worker's compensation insurance on his employees, who is responsible for providing the insurance?

    a. Owner

    b. General contractor

    c. State government

    d. No one

4. Which of the following financial statements indicates the financial condition of a company at a given point in time?

    a. Income statement

    b. Cash flow statement

    c. Balance sheet

    d. Funds flow statement

Notes

5. A first aid kit is checked before it is sent to the job and is then checked:

   a. Every week
   b. Biweekly
   c. Monthly
   d. Quarterly

6. Which of the following would be classified as a long-term debt?

   a. Current annuity of long-term debt
   b. Mortgage on a company office building
   c. Accounts payable
   d. Accumulated depreciation

7. What is another name for owner's equity?

   a. Operating capital
   b. Liquid assets
   c. Net worth
   d. Liability

8. Which of the following statements defines *net working capital*?

   a. The amount that current assets exceed the current liabilities
   b. The amount listed as an item overhead expense
   c. The difference between direct cost and the cost of sales
   d. Current liabilities less overhead expense

9. When two companies enter into a project as a joint venture, what must occur if one company does not have a contractor's license?

   a. The company with the license will manage
   b. A license in the name of the joint venture shall be obtained
   c. This is permitted as long as at least one firm has a license
   d. Other

10. Based on an accrual basis of accounting, earnings that have not been received will be displayed as:

    a. Current assets
    b. Fixed assets
    c. Current, liabilities
    d. Owner's equities

11. The required quality of a project is described in the ___________ .

    a. budget
    b. specifications
    c. plans
    d. general conditions

12. Current assets are assets that can be converted into cash within:

    a. 1 year
    b. 6 months
    c. 1 month
    d. 2 years

13. A form of business organization that has a tax structure similar to a partnership but is organized as a corporation is a __________ .

    a. limited partnership
    b. subchapter s corporation
    c. general partnership
    d. nonprofit organization

Study Tip

A qualifying agent is the person taking the exam. Each state has different requirements but most require the person to be employed full time and be at least 18 years of age. Make sure you understand your state's requirements.

Study Tip

Be sure to understand the amount of time allowed to replace a qualifying agent in the event he or she leaves the company.

14. Regarding contracts, which of the following statements is *not true*?

    a. Words prevail over numbers
    b. General clauses prevail over special clauses
    c. When properly documented, handwritten agreements prevail over typed provisions
    d. A contract deemed ambiguous is interpreted against the party that provided it

15. Which of the following work-related injuries does *not* require the employer to report to OSHA?

    a. A broken wrist
    b. An x-ray
    c. Cancer resulting from long-term exposure to work-related chemicals
    d. A sprain that removed an employee from work for 2 days

16. What is a benefit of good inventory control?

    a. Better for safety
    b. Better preparation for increase in business
    c. Reduced time and expense in handling materials
    d. Ability to ship faster to off-site storage

17. Periodic payments made by the owner to the contractor over a long project are called:

    a. Progress payments
    b. Acceleration payments
    c. Performance payments
    d. Provisional payments

18. What accounting method records income when cash is received?

    a. Accrual basis
    b. Cash basis
    c. Single entry
    d. Double entry

19. What portion of the worker's compensation insurance premium can an employer charge to an employee?

    a. A maximum of 2% of the employee's weekly gross earnings
    b. No deduction is allowed
    c. Up to 50% of the premium
    d. A flat fee of $25 per week

20. If an employee working for a corporation and the licensing examination for the corporation, who owns the license when it is issued?

    a. The employee
    b. The stockholders
    c. The president of the corporation
    d. The corporation itself

21. An employer must maintain records about work-related injuries for _______ year(s).

    a. 3
    b. 5
    c. 1
    d. 7

Study Tip

It is important to know the requirements of your licensing board with regard to change of address. Most boards allow 30 days for you to properly notify them after you move your office of record.

Study Tip

Does your state have a recovery fund? If so, you will need to understand the fund limitations and requirements.

22. Which of the following statements is *not true* regarding a bar chart method of scheduling?

    a. It shows the start and finish times for each task
    b. It does not show how tasks relate to each other
    c. It is most common scheduling system found on the job site
    d. All are true

23. When a clause in a contract calls for liquidation damages, it is referring to:

    a. A payment per day for finishing after the contract deadline
    b. A payment due to flooding
    c. An expedient payment for improper work
    d. None of the above

24. Which of the following statements is *true* about partnerships?

    a. Each partner may cause the other to be jointly liable for any action incurred during normal business
    b. Each partner is personally liable for his actions only
    c. Each partner may sell his share of the partnership without the consent of the other partners
    d. A general partner may discharge himself from all liability by declaring himself to be a limited partner

25. Which of the following acts requires payment of prevailing wage rates and fringe benefits on federally assisted construction?

    a. Davis-Bacon Act
    b. Walsh-Healey Public Contracts Act
    c. Wage Garnishment Law
    d. Fair Labor Standards Act

PART SIXTEEN

OSHA

Questions are based on safety issues as they relate to OSHA.

The following section is based on OSHA 1926, Construction Industry Regulations. If the OSHA book is an approved reference material listed for your exam, these questions represent the types of questions that will be covered.

Questions in this quiz cover the following subject matter:

- Scaffolding
- Ladders
- Fall protection
- Excavation
- Storage
- Tools

Study Tip

You may find the Table of Contents to be more helpful than the Index when using the OSHA book. While both are helpful, you will sometimes be forced to head to the general area of the subject you are searching.

OSHA Practice Exam 1

1. The maximum slope for an 8 feet excavation in Type B soil is:

   a. ¾:1

   b. ½:1

   c. 1:1

   d. 2:1

2. The minimum perpendicular clearance between fixed ladder rungs, cleats, and steps and any obstruction behind the ladder is ______ inches.

   a. 6

   b. 7

   c. 8

   d. 9

3. The use of electrical cords for hoisting or lowering shall:

   a. Be allowed

   b. Not be allowed

   c. Only be permitted during daylight hours

   d. Not be permitted during dusk or evening hours

4. When necessary to require employees to work in a trench, the excavated materials shall be placed a minimum of ____ away from the edge of the trench.

   a. 0'

   b. 1'

   c. 2'

   d. 3'

5. Eye protection near dangerous working conditions is:

   a. Required at the employee's cost.
   b. Required at the employer's cost.
   c. Not required, except by union regulations
   d. Not required

6. Who is responsible for the fire safety and prevention on a job site through completion of a project?

   a. OSHA
   b. Employer
   c. Fire marshall
   d. Building inspector

7. When employees are required to be in trenches of 4 feet deep or more, an adequate means of exit such as a ladder or steps shall be provided and located so as to require no more than ___ feet of lateral travel.

   a. 10
   b. 15
   c. 25
   d. 30

8. How much water is allowed in an excavation before work must cease?

   a. 3" or less at the edge
   b. 2" or less at the edge
   c. 1" or less at the edge
   d. None

Study Tip

When employer and employee is an option, employer will be the correct answer nearly every time.

Study Tip

Don't forget to highlight each of the answers as they are located. Never guess on an answer when working through the practice exams. Remember you are learning to use your reference books.

9. The minimum illumination in active general construction storage areas is:

   a. 3 footcandles
   b. 5 footcandles
   c. 10 footcandles
   d. 30 footcandles

10. After being sent to the job site, a first aid kit is checked _______ .

   a. weekly
   b. biweekly
   c. monthly
   d. quarterly

11. OSHA requires a ladder in every excavation exceeding:

   a. 5 feet
   b. 3 feet
   c. 4 feet
   d. 6 feet

12. Compressed air shall not be used for cleaning purposes, except where reduced to less than _______ psi.

   a. 10
   b. 20
   c. 30
   d. 70

13. Sand is considered what soil type?

   a. Type A
   b. Type B
   c. Type C
   d. Type D

14. According to OSHA, when employees are exposed to inhalation, ingestion, skin absorption, or contact with any substance at concentrations above the __________ , they shall be evacuated unless specific forms of personnel protective equipment or devices are provided.

    a. threshold limit value

    b. PPM mandate

    c. PEL levels

    d. oxygen deprivation level

15. According to OSHA, the height at which fall protection is required is ____ feet.

    a. 6

    b. 8

    c. 10

    d. 12

16. Circular saws shall be equipped with a _________ switch.

    a. position on-off

    b. momentary on-off

    c. lock-on control

    d. constant pressure

17. Combustible debris should be removed from a construction job site ______________________________________ .

    a. when the job is complete

    b. only in the event of hazardous fire conditions

    c. periodically by any means

    d. at regular intervals

Study Tip

Fall protection for scaffolding is different than general fall protection. Make sure you know both.

Study Tip

The employer is always responsible for providing Personal Protective Equipment in the event the employee does show up with adequate protection. "I told him" is not a defense.

18. When using a stepladder:

    a. The top step can be used

    b. The cross bracing on the back can be used as a step

    c. Both hands must be used on the sides

    d. It should be inspected by a competent person for visible defects on a periodic basis

19. When employees are required to be in trenches of 4 feet deep or more, an adequate means of egress shall be provided and located so as to require not more lateral travel than ______ feet.

    a. 10

    b. 12

    c. 15

    d. 25

20. The maximum allowable slope for excavations less than 20 feet in depth for Type C soil is:

    a. ¾ horizontal to 1 vertical

    b. 1 horizontal to 1 vertical

    c. 1½ horizontal to 1 vertical

    d. 2 horizontal to 1 vertical

21. A 10-foot-deep trench is to be dug in granular sand. What is the maximum allowable bank slope?

    a. 1:1

    b. 1½:1

    c. ½:1

    d. ¼:1

22. When using a portable ladder for reaching an elevated surface or for ingress or egress of an excavation, the ladder must extend to what minimum height above the surface?

    a. 2.5 feet

    b. 3 feet

    c. 3.5 feet

    d. 4 feet

23. ______ soil has a maximum allowable slope of 1:1.

    a. Type A

    b. Type B

    c. Type C

    d. Type D

24. Each of the following is a soil type *except*:

    a. Gravel

    b. Sediment

    c. Sand

    d. Clay

25. OSHA requires the employer to provide ________________ for all employees/workers in all operations where there is an exposure to hazardous conditions.

    a. no-smoking areas

    b. an opportunity to change positions

    c. new tools

    d. hardhats

Study Tip

Be sure to tab the location that explains types of soils and allowable slopes.

Study Tip

Understanding how the OSHA book is laid out will be crucial to your success on exam day to locate answers for difficult questions. Consider tabbing major sections such as: Scaffolding, Excavations, Ladders etc.

OSHA Practice Exam 2

1. What is the *maximum* amount of hours an employee can work if exposed to a D.B.A level of 102?

   a. 1 hour
   b. 1.5 hours
   c. 2 hours
   d. 3 hours

2. When intermediate vertical members such as balusters used between posts are positioned, the space between them shall not exceed ____ inches.

   a. 10
   b. 15
   c. 19
   d. 20

3. Combustible material stored outdoors shall be at least how many feet away from the structure?

   a. 20 feet
   b. 50 feet
   c. 10 feet
   d. 75 feet

4. How often should powder-actuated tools be tested?

   a. Once a week
   b. Per manufacturer's requirements
   c. Each day after use
   d. Each day before use

5. Safety belt lanyard hardware shall be able to withstand a tensile loading strength of up to _____ pounds before cracking, breaking, or deforming.

   a. 1,000
   b. 3,000
   c. 4,000
   d. 5,000

6. Hooks on an attachable ladder shall be positioned so that the bottom rung is not more than ____ inches above the scaffold supporting level.

   a. 10
   b. 18
   c. 20
   d. 24

7. Stacks of lumber that are to be handled manually shall not be stacked more than ____ feet.

   a. 10
   b. 15
   c. 16
   d. 20

8. When debris is dropped through holes in the floor without the use of chutes, the area onto which the material dropped shall be completely enclosed with barricades and located not less than ______ feet back from the projected edge of the opening above.

   a. 2
   b. 4
   c. 6
   d. 8

Study Tip

Be sure to read each section carefully as you identify your answers.

Study Tip

OSHA questions resemble Code questions and often have call for MAXIMUM, MINIMUM, MUST BE, SHALL BE and REQUIRED. To distinguish, simply ask yourself if the question is addressing safety of the jobsite (OSHA) or safety that will directly impact the inhabitants of the structure (CODE).

9. All equipment used in site clearing shall be equipped with a roll-over canopy structure of not less than ___ steel plate or ____ wire mesh with openings of no more than 1 inch.

   a. ¼"

   b. ⅛"–¼"

   c. ¼"–⅛"

   d. ½"–¾"

10. Integral prefabricated scaffold access frames shall be provided with rest platforms at ____ foot intervals.

    a. 15

    b. 25

    c. 35

    d. 45

11. What is the required rating for fire extinguishers placed on job sites where more than 5 gallons of flammable liquids are used?

    a. 5A

    b. 5B

    c. 10A

    d. 10B

12. A safety belt lanyard shall have a length allowing a maximum fall of ___ feet.

    a. 2

    b. 4

    c. 6

    d. 8

13. When bricks are stored on a job site, the maximum height they can be stacked without additional requirements is ____ feet.

    a. 5
    b. 6
    c. 7
    d. 8

14. How often should first aid kits be checked?

    a. Monthly
    b. Weekly
    c. Biweekly
    d. Daily

15. What is the minimum diameter of a safety lanyard?

    a. ¼ inch
    b. ⅜ inch
    c. ½ inch
    d. ¾ inch

16. Fiber rope slings shall have a minimum clear length of rope between eye splices equal to ____ times the rope diameter.

    a. 5
    b. 10
    c. 15
    d. 20

Study Tip

Typically, OSHA questions represent a small portion of any state exam.

Study Tip

Do not spend more than 5 minutes looking for an answer. Select your best guess, mark the question and move on. If you have time, come back to the question.

17. All handheld circular saws shall be equipped with a:

    a. Positive on-off switch
    b. Momentary contact switch
    c. Constant pressure switch
    d. None of the above

18. When dropping waste more than ____ feet, an enclosed chute of wood or an equivalent shall be used.

    a. 10
    b. 15
    c. 20
    d. 25

19. Safety nets shall be provided when workplaces are more than ____ feet above the ground and no other protection is provided.

    a. 10
    b. 15
    c. 20
    d. 25

20. A ladder jack scaffold platform shall not exceed ____ feet in height.

    a. 10
    b. 15
    c. 20
    d. 25

21. A fixed guardrail system, used to protect an opening must be capable of sustaining the pressure exerted by a ____ person if there is a chance of someone falling through it.

    a. 150 lb.

    b. 175 lb.

    c. 200 lb.

    d. 225 lb.

22. Safety nets must extend beyond the work surface by at least ____ feet.

    a. 4

    b. 6

    c. 8

    d. 10

23. The minimum distance between a scaffold and a 300-volt power insulated line is ___ feet.

    a. 3

    b. 6

    c. 9

    d. 10

24. When using a ladder, it must extend at least ___ feet above the surface to which it is providing access.

    a. 1

    b. 2

    c. 3

    d. 4

Study Tip

Do not run out of time when you are taking your exam. Always be aware of the time left. Never allow the clock to run out without at least selecting answers to the remaining questions.

Study Tip

Remember to stay calm throughout your exam. Becoming frustrated will only make it worse.

25. Stairway-type ladders shall be provided with a rest platform at ____ foot *maximum* vertical intervals.

    a. 8

    b. 10

    c. 12

    d. 14

PART SEVENTEEN

Comprehensive Final

Questions are based on general knowledge of concrete.

The following comprehensive final is based on questions from the preceding chapters. This exercise will prove most beneficial to exam candidates that are preparing to take an exam with multiple approved reference books. By taking questions from a wide variety of subjects, the importance of reading the question, deciding what chapter of the code to locate, or whether or not the question can be answered in the code book becomes apparent.

Certain words will hint that the question can be answered from the code book. These words include:

- Minimum
- Maximum
- Must be
- Shall be
- Required

Questions that are based on theory and general knowledge can be identified by words such as:

- Recommended
- Suggested
- What is used
- Typically

Notes

1. Utility-grade studs cannot be spaced more than ______ inches on center.

   a. 12

   b. 16

   c. 20

   d. 24

2. Foundation plates or sills should be attached to concrete foundation with bolts:

   a. ½ inches in diameter with a penetration of at least 7 inches

   b. ¾ inches in diameter with a penetration of at least 7 inches

   c. ½ inches in diameter with a penetration of at least 6 inches

   d. ¾ inches in diameter with a penetration of at least 6 inches

3. The type of mortar best suited for use below grade is:

   a. Type K

   b. Type M

   c. Type N

   d. Type S

4. The building permit must be:

   a. Displayed in the contractor's office

   b. Posted on the job site

   c. Provided to the inspector when requested

   d. Posted in the building official's office

5. A brick cut lengthwise across the end is called a _______ closure.

   a. queen
   b. king
   c. bat or half
   d. split

6. The best source for projecting future project and company overhead percentage rates is ____________________ .

   a. historical data
   b. bid documents
   c. unit pricing
   d. job cost records

7. Exterior air intake shall not be located in any of the following areas *except*:

   a. A garage
   b. Outside a dwelling
   c. A basement
   d. An attic

8. Which of the following is a consideration when applying tile?

   a. Climate conditions
   b. Roof slope
   c. Type of tile
   d. All of the above

Notes

Notes

9. The use of electrical cords for hoisting or lowering shall:

   a. Be allowed

   b. Not be allowed

   c. Only be permitted during daylight hours

   d. Not be permitted during dusk or evening hours

10. Normal concrete is designed to reach its intended compressive strength within ______ days.

    a. 7

    b. 14

    c. 21

    d. 28

11. For ceilings not attached to rafters with a roof live load of 20 psf and a dead load of 10 psf, the maximum span allowed for a 2 × 8 rafter to be spaced 16 inches on center is ______. *Note:* Material is SYP #2.

    a. 18 feet 6 inches

    b. 19 feet 5 inches

    c. 14 feet 4 inches

    d. 19 feet 3 inches

12. What is the minimum depth of a masonry chimney firebox?

    a. 20 inches

    b. 24 inches

    c. 36 inches

    d. 301

13. The minimum uniformly distributed live load for a room with an intended use of sleeping is designated as _______ psf.

    a. 10
    b. 20
    c. 30
    d. 50

14. For the purpose of the code, a vehicle covering is considered a carport if it is open on:

    a. At least one side other than the entry
    b. All four sides
    c. At least two sides
    d. At least one side and not adjacent to the structure

15. For each inch of reduced slump in concrete, the pounds per square inch is reduced by:

    a. 100 psi
    b. 125 psi
    c. 150 psi
    d. 200 psi

16. What is another name for owner's equity?

    a. Operating capital
    b. Liquid assets
    c. Net worth
    d. Liability

Notes

Notes

17. The net free ventilation area for attics must be at least _______ of the area. *Note:* No exceptions.

    a. 1/150

    b. 1/200

    c. 1/250

    d. 1/300

18. Wood shakes shall be laid with a minimum side lap not less than _______ inches.

    a. 3/4

    b. 1

    c. 1 1/4

    d. 1 1/2

19. Calculate the common number of rafters for a building that is 24 feet long with a 1 foot overhang. The slope factor for the building is 1.118.

    a. 14

    b. 15

    c. 16

    d. 17

20. Sand is considered to be what soil type?

    a. Type A

    b. Type B

    c. Type C

    d. Type D

21. A plain masonry hollow wall that is 7 feet tall with 6 feet of unbalanced backfill should be at least _______ inches thick. The soil type is GW.

    a. 6

    b. 8

    c. 10

    d. 12

22. Chimney terminations shall not be less than _______ feet above the highest point where the chimney passes through the roof.

    a. 2

    b. 3

    c. 6

    d. 10

23. A permit is deemed void if work does not begin within _______ days of issuance.

    a. 30

    b. 60

    c. 90

    d. 180

24. The minimum diameter head permitted to be on a fastener for asphalt shingles is _______ inches?

    a. 0.105

    b. 0.375

    c. 0.50

    d. 0.75

Notes

Notes

25. A smoke-developed index of no more than ______ should be maintained for all wall and ceiling finishes.

    a. 200

    b. 250

    c. 375

    d. 450

26. Beams and girders must bear at least ______ inch(es) on a masonry column.

    a. 1

    b. 1½

    c. 1¾

    d. 3

27. Vapor retarder is not required between the base course/subgrade and a concrete floor slab in which of the following circumstances?

    a. Typical driveway and walkways

    b. Unheated accessory structures

    c. For installations approved by the building official based on local site conditions

    d. All of the above

28. Purlins used to support roof loads must be at least:

    a. 2 × 6

    b. One size larger than the rafter it is bracing

    c. No less than one size smaller than the rafter it is bracing

    d. The size of the rafter it is supporting

29. *Given:* A carpenter drops a hammer from a roof and injures a curious bystander. Which type of insurance would cover the victim's medical costs?

    a. Builder's risk

    b. Named-peril risk

    c. Liability

    d. Risk management

30. How many studs will be needed if a wall is 44 feet long and the spacing is 16 inches on center?

    a. 33

    b. 34

    c. 35

    d. 36

31. What is the nominal dimension of a standard concrete masonry unit (Block)?

    a. 7⅝" × 7⅝" × 16"

    b. 7⅝" × 8" × 15⅝"

    c. 8" × 7⅝" × 15⅝"

    d. 8" × 8" × 16"

32. SPF #2 is to be installed 16 inches on center in a residential living area with a dead load of 20 psf. The length of the room is 14 feet 6 inches. What size floor joist is required?

    a. 2 × 6

    b. 2 × 8

    c. 2 × 10

    d. 2 × 12

Notes

Notes

33. What is the maximum stud spacing if particleboard wall sheathing is to be nailed to the stud walls? *Note:* M-2 Exterior glue grade, ½ inch thickness.

    a. 12 inches on center

    b. 16 inches on center

    c. 24 inches on center

    d. 30 inches on center

34. Particleboard (M-1/M-S grade) wall sheathing nailed to 16 inches on center studs must be at least _______ inch(es) in thickness.

    a. ½

    b. ⅜

    c. ⅝

    d. 1

35. The most common welded-wire fabric used for residential applications has wires spaced _______ inches apart in each direction.

    a. 2

    b. 4

    c. 6

    d. 8

36. Masonry chimneys shall be lined appropriately for the type of appliance connected. Residential-type appliances having a clay flue lining that must comply with the requirements of:

    a. ASTM C 27

    b. ASTM C 199

    c. ASTM C 315

    d. ASTM 1261

37. What is the maximum notch depth allowed for a 2 × 12 nominal floor joist? *Note:* The joist spans 14 feet 6 inches and the notch needs to be made 5 feet from one end.

    a. 1⅞"

    b. 1¾"

    c. 2"

    d. None of the above

38. When single top plates are permitted, the plate must be adequately tied at joints, corners, and intersecting walls by a galvanized steel plate that is nailed to each wall or segment by _______ nails on each side.

    a. Eight 16d

    b. Four 16d

    c. Six 8d

    d. Four 8d

39. A 2 × 6 SYP #2 material is to be used for the ceiling joists with a live load of 10 psf. The joists, spaced 24 inches on center, will be installed to support an attic that will not be used for storage. What is the maximum span?

    a. 11 feet 0 inches

    b. 15 feet 6 inches

    c. 15 feet 11 inches

    d. 17 feet 8 inches

40. Coal-tar saturated organic felt used for built-up roof cover must conform to what standard?

    a. ASTM D 227

    b. ASTM D 450

    c. ASTM D 43

    d. ASTM D 4022

Notes

Notes

41. *Given:* The floor of a combustion chamber to the top of the flue is 18 feet. The fireplace opening is 2,200 square inches. What is the minimum cross-sectional area required for a retangular chimney flue?

    a. 125 square inches
    b. 140 square inches
    c. 168 square inches
    d. 214 square inches

42. Hydrated lime that has been formed into putty before packaging is called __________.

    a. mixed
    b. prehydrated
    c. slaked
    d. preserved

43. ______ soil has a maximum allowable slope of 1½:1.

    a. Type A
    b. Type B
    c. Type C
    d. Type D

44. For an emergency escape opening in a habitable space, the minimum net clear opening width is specified as ______ inches.

    a. 20
    b. 22
    c. 24
    d. 28

45. Handrail height for stairs, when required, should be a minimum ______ inches and a maximum ______ inches, measured vertically from the sloped plane adjoining the tread nosing.

    a. 38/34 (respectively)

    b. 34/36 (respectively)

    c. 34/38 (respectively)

    d. 36/38 (respectively)

46. Gravel or crushed stone drains installed for the purpose of foundation drainage shall extend at least ______ beyond the outside edge of the footing.

    a. 0.5 feet

    b. 1 foot

    c. 1.5 feet

    d. 2 feet

47. A one-story building with a width of 28 feet requires ______ girder(s). *Note:* The span is 10 feet 2 inches.

    a. Four 2 × 12

    b. Four 2 × 10

    c. Four 2 × 8

    d. Three 2 × 10

48. Fasteners for attaching structural sheathing (shear panels) to steel wall framing must be installed with a minimum edge distance of ______ inch.

    a. ¼

    b. ½

    c. ⅜

    d. 1

Notes

Notes

49. What species of lumber must be used if 2 × 6 nominal ceiling joists are to span 14 feet 8 inches and be spaced 16 inches on center? The live load is 20 psf, and the application is uninhabitable attics with limited storage.

 a. Douglas fir-larch

 b. Hem-fir

 c. Southern pine

 d. Spruce-pine-fir

50. All of the following fasteners are used for asphalt shingles *except:*

 a. Galvanized steel

 b. Copper

 c. Aluminum

 d. Nickel

51. Based on a brick factor of 6.5 per square foot, how many bricks will be needed for a wall that is 35 feet long and 12 feet tall?

 a. 2,730

 b. 2,600

 c. 2,450

 d. 2,340

52. Footings for masonry chimneys in areas not subjected to freezing shall be at least _______ inches below finished grade.

 a. 6

 b. 12

 c. 18

 d. 24

53. Chimneys located inside a building shall have minimum airspace combustibles of _______ inches.

    a. 2

    b. 4

    c. 6

    d. 12

54. The space between the end of the rebar and the form is referred to as _______ .

    a. cover

    b. bond

    c. space

    d. void

55. _________ units are reduced-thickness units, or fire clay, used in veneer applications.

    a. Glazed brick

    b. Thin brick veneer

    c. Green bricks

    d. Paving bricks

56. ABC Contractors Company has 21 employees. According to OSHA, this contractor must keep a record for _______ years after each recorded accident while on the job site for each one of its employees.

    a. 3

    b. 4

    c. 5

    d. 6

Notes

Notes

57. Combustible debris should be removed from a construction job site ________________________________.

    a. when the job is complete
    b. only in the event of hazardous fire conditions
    c. periodically and by any means
    d. at regular intervals

58. Untreated wood joists should not be installed closer than ______ inches.

    a. 12
    b. 14
    c. 16
    d. 18

59. Where masonry veneer is used, concrete foundation walls are required to extend ______ inches above the finished grade adjacent to the foundation walls at all points.

    a. 2
    b. 4
    c. 6
    d. 8

60. Sheathing with a span rating of 24/0 to be used for a subfloor can be installed with a maximum span of ______ inches.

    a. 0
    b. 16
    c. 24
    d. 32

61. Buildings must be provided with exterior and interior braced wall lines. The spacing of the braced wall lines shall not exceed _______ feet in both the longitudinal and transverse directions in each story. *Note:* No exceptions apply.

    a. 10
    b. 25
    c. 35
    d. 50

62. The type of nails that can be substituted for 16d nails when fastening rafter connections is:

    a. 10d
    b. 12d
    c. 40d box
    d. No substitutions provided by code

63. On roofs sloped 4:12 or greater, underlayment shall be at least one layer applied shingle fashion, starting from the eaves and lapped _______ inch(es).

    a. 1
    b. 2
    c. 4
    d. 5

64. What is the absolute minimum thickness of a heart extension?

    a. ⅜"
    b. 2"
    c. 4"
    d. 8"

Notes

Notes

65. With regard to masonry units, a *nominal* dimension refers to:

    a. Actual dimension

    b. Standard brick sizes

    c. Nominal brick plus the mortar joint

    d. Actual brick size plus the mortar joint

66. A 10 feet × 10 feet habitable room should have an aggregate glazed area of ______ square feet to provide natural lighting.

    a. 5

    b. 6

    c. 7

    d. 8

67. Pier and curtain wall foundations are permitted to support light frame construction of not more than two stories. All of the following requirements must be met by code *except:*

    a. The load-bearing wall shall be placed on continuous concrete footings that are placed integrally with the exterior wall

    b. The thickness of the load-bearing wall shall not be less than 3⅜" nominal

    c. The maximum height of a 4" load-bearing masonry foundation wall supporting wood framed walls shall be more than 4' in height

    d. The unbalanced fill for 4" foundation walls shall not exceed 24" for solid masonry.

68. A hold to be bored into a floor joist for the purpose of running a plumbing pipe or electrical conduit:

    a. Must not exceed one third of the depth of the joist

    b. Must not exceed one sixth of the depth of the joist

    c. Cannot be bored in the middle third of the joist

    d. Is not allowed unless approved by a qualified engineer

69. Interior partitions (nonbearing) may be constructed with at least a _______ top plate.

    a. double top plate
    b. single top plate
    c. double top plate, if wall height exceeds 10 feet
    d. None of the above

70. What is the maximum span of a 2 × 6 SPF #2 ceiling joist to be spaced 12 inches on center? The attic will not be utilized in any way. Dead load is 5 psf, and the live load is 10 psf.

    a. 18 feet 8 inches
    b. 14 feet 9 inches
    c. 16 feet 3 inches
    d. 13 feet 9 inches

71. Mineral-surfaced roll roofing shall not be applied on roof slopes with less than _______ slope.

    a. 2%
    b. 8%
    c. 12%
    d. 33%

72. The upper edge of a chimney cleanout shall be located at least _______ inches below the lowest chimney inlet opening.

    a. 2
    b. 4
    c. 6
    d. 8

Notes

Notes

73. Concrete is made up of how many parts?

    a. 1

    b. 2

    c. 3

    d. 4

74. How many 8 inches × 16 inches CMUs will be needed for a wall that is 45 feet long and 12 feet tall. The total square feet of openings is 200.

    a. 340

    b. 383

    c. 540

    d. 608

75. A ______ masonry wall is constructed with two wythes that can react independently of one another and are separated by a continuous airspace of at least 2 inch (typical).

    a. veneer

    b. hollow

    c. cavity

    d. composite

76. Which of the following statements is *true* regarding site plans submitted for demolition projects?

    a. Site plans are not required for structures to be removed

    b. Site plans must show the construction to be demolished

    c. Site plans must show the size and location of existing construction that will remain

    d. Both b and c

77. A guardrail system is required for protection when a porch or balcony is located more than _______ inches above the grade or floor below.

    a. 24
    b. 30
    c. 36
    d. 40

78. What size and type of gypsum board is required to separate a garage from a habitable room located above it?

    a. ½ inch Type X
    b. ⅝ inch Type X
    c. ¾ inch Type X
    d. 3⁄16 inch Type X

79. All permanent supports are not required to be protected from frost. Which of the following is required to have frost protection?

    a. A freestanding building of more than 400 square feet
    b. A freestanding building of 400 square feet or less with an eave 10 feet in height.
    c. A freestanding building 400 square feet or less with an eave 8 feet in height
    d. A freestanding building of 300 square feet with an eave 8 feet in height

80. In residential structures, when there is usable space above and below the concealed space of a floor/ceiling assembly, draftstops must be installed so that the area of the concealed space does not exceed _______ square feet.

    a. 500
    b. 800
    c. 1,000
    d. 1,200

Notes

Notes

81. When joists exceed a nominal 2 inches × 12 inches (6:1 depth-to-thickness ratio), bridging must be installed for each ______ feet of span.

    a. 4
    b. 8
    c. 12
    d. 16

82. What is the maximum spacing of a 2 × 6 for a bearing wall to support a roof and ceiling only?

    a. 16 inches on center
    b. 20 inches on center
    c. 24 inches on center
    d. 28 inches on center

83. A notch must be placed in a ceiling joist. The nominal size of the joist is 2 × 10 and spans 12 feet 8 inches. What is the maximum notch that can be placed on the top, 2 feet from the exterior wall?

    a. 1.67"
    b. 1.54"
    c. 1.33"
    d. 0

84. Clay tile shall be installed on a roof with a minimum slope of:

    a. 2½:12
    b. 3:12
    c. 4:12
    d. 5:12

85. When employees are required to be in trenches of 4 feet deep or more, an adequate means of exit such as a ladder or steps shall be provided and located so as to require no more than ______ feet of lateral travel.

    a. 10
    b. 15
    c. 25
    d. 30

86. Footings for masonry fireplaces shall extend at least ______ inches beyond the face of the fireplace.

    a. 4
    b. 6
    c. 10
    d. 12

87. The most common type of reinforcement used for the installation of footings is _____________ .

    a. deformed steel
    b. smooth steel
    c. rigid steel
    d. #3 smooth steel

88. Which of the following types of brick bond patterns would be considered the weakest?

    a. Flemish
    b. Running
    c. Stack
    d. English

Notes

Notes

89. For ramps intended for use other than specifically for means of egress, the maximum allowable slope is ______ .

    a. 8%

    b. 10%

    c. 10.5%

    d. 12.5%

90. Calculate the board feet of lumber for 12 studs (2 × 4) that are 9 feet in length.

    a. 72 board feet

    b. 78 board feet

    c. 80 board feet

    d. 97 board feet

91. All rescue openings are required to have at least a net clear opening of ______ square feet unless situated on a grade level.

    a. 5.7

    b. 5.0

    c. 4.7

    d. None of the above

92. Joist framing from opposite sides of a beam or girder shall be lapped at least ______ inches.

    a. 2

    b. 3

    c. 4

    d. 6

93. A hole may not be bored closer than _______ inch to the edge of a stud located in an exterior load-bearing framed wall.

    a. ¼
    b. ½
    c. ¾
    d. ⅝

94. A hole must be bored in a ceiling joist and needs to be placed near the edge. What is the closest the hole can be placed to the edge?

    a. 1"
    b. 2"
    c. 2½"
    d. 3"

95. Current assets are assets that can be converted into cash within:

    a. 1 year
    b. 6 months
    c. 1 month
    d. 2 years

96. Which of the following types of wood shakes is *not* approved for grade No. 1?

    a. 18" shakes of naturally durable wood with 7.5" exposure
    b. 24" taper-sawn shakes of naturally durable wood with 7.5" exposure
    c. 24" taper-sawn shakes of naturally durable wood with 10" exposure
    d. 18" preservative-treated taper shakes of Southern yellow pine with 7.5" exposure

Notes

Notes

97. A chimney shall not be corbelled more than ______ of the chimney's wall thickness from a wall.

    a. 0.25

    b. 0.375

    c. 0.5

    d. 0.75

98. Typical rebar comes in ______ foot lengths that can be cut or bent on the job site according to needs specified.

    a. 10

    b. 15

    c. 20

    d. 25

99. What grade of concrete brick is required when the structure will be exposed to severe frost?

    a. Grade N

    b. Grade S

    c. Grade M

    d. Grade I

100. A sprayed polyurethane foam roof requires a protective coating to be applied no later than ______ after foam application.

    a. 2 hours

    b. 12 hours

    c. 72 hours

    d. can be applied immediately

PART EIGHTEEN

Answer Keys

Notes

Part 1: Math Concepts Answer Key

Converting Inches to Feet Answers/Solutions:

1. (c) 14 ÷ 12 = 1.16666666. Round by identifying the rounding digit as the second number. The digit to the right of the rounding digit is greater than 5; therefore, the rounding digit is increased by 1, and the remaining digits are dropped.
2. (b) 8 ÷ 12 = 0.6666666. Round by identifying the rounding digit as the second number. The digit to the right of the rounding digit is greater than 5; therefore, the rounding digit is increased by 1, and the remaining digits are dropped.
3. (a) 3 ÷ 12 = 0.25. Round by identifying the rounding digit as the second number. There are no numbers to the right of the rounding digit, so you are finished.

Applying the Concepts of Rounding and Converting Inches to Feet Answers/Solutions:

1. (b) 8 ÷ 12 = 0.66666. Round by identifying the rounding digit as the second number. The digit to the right of the rounding digit is greater than 5; therefore, the rounding digit is increased by 1, and the remaining digits are dropped.
2. (b) 14 ÷ 12 = 1.16666666. Round by identifying the rounding digit as the second number. The digit to the right of the rounding digit is greater than 5; therefore, the rounding digit is increased by 1, and the remaining digits are dropped. The total of 14" is derived from adding the space to support a 4" brick, an 8" block, and the 2" of air separating the two units.
3. (c) 21 ÷ 12 = 1.75. Round by identifying the rounding digit as the second number. There are no numbers to the right of the rounding digit, so you are finished. The 21" were derived from the information provided. The dimension (to the right) from the grade to the bottom of the slab is denoted as 24". The slab thickness is 5". Add the two together for a total of 29". Next, subtract the 8" brick ledge. The remainder is 21".
4. (b) 29 ÷ 12 = 2.416666.
5. (d) 14.67 − 1.17 − 1.17 = 12.33. Begin by converting inches to feet. 14'8" is converted to 14.67 after dividing the 8 inches by 12 to equal 0.67 (don't forget to round to the nearest hundredth). Next, determine the thickness of the wall by adding the brick, block and airspace. This totals 14 inches. Now convert the 14 inches to feet by diving by 12. 14 divided by 12 = 1.1666666. Rounded to the

nearest hundredth, the thickness of the wall is 1.17. Now, subtract the thicknessof the wall from the overall dimension of the building. 14.67 – 1.17 – 1.17 = 12.33

6. (b) 14.67 – 0.5 – 0.5 = 13.67. Begin by converting inches to feet. Subtract the brick and airspace on each side from the overall dimension of the wall.
7. (c) 23 – 1.17 – 1.17 = 20.66. Subtract the thickness of each wall from the length of the building.

Using Area (Square Feet) to Estimate Sheathing, Drywall, Brick and Block Answers/Solutions:

1. (c) Two walls measure 24' × 8' for a total of 384 square feet. Two walls measure 30' × 8' for a total of 480 square feet. The ceiling measures 24' × 30' for a total of 720 square feet. Add each of these together for a total of 1,584 square feet. The total square feet divided by 32 (which is the square feet of a 4' × 8' piece of drywall) equals 49.5. Round this to 50.
2. (a) 2,875 square feet divided by the area of the material to be installed of 32 square feet equals 89.84. Round to 90 (closest answer and the quantity to order).
3. (b) Half-inch gypsum is specified for walls only. Two walls measure 12' × 8' for a total of 192 square feet. Two walls measure 14' × 8' for a total of 224 square feet. The two totals added together equal 416 square feet. 416 square feet divided by 28 square feet (size of the material) equals 8.67. Round to 9.
4. (b) Two walls measure 20' × 18' for a total of 720 square feet. Two walls measure 32' × 18' for a total of 1,152 square feet. Add each of these together for a total of 2,106 square feet. The total square feet multiplied by 1.125 equals 2,106.
5. (c) Determine the brick multiplier by calculating 7 × 2.75 = 19.25 and dividing this into 144 square inches. The multiplier is 7.4805. The square feet of the area is 10 × 20 = 200 + 10% (waste) for a total of 220 square feet: 220 × 7.4805 = 1,645.71, which rounds to 1,646.
6. (a) 32 × 8 = 256 square feet of wall area. Subtract the 28 square feet of openings = 228 square feet. Determine the block by multiplying by 1.125 = 256.5, which rounds to 257. This is the only choice with 257 block; therefore, calculating brick is not necessary. To determine the quantity of brick, begin with determining the multiplier: 2.5 × 8 = 20 inches. Divide 144 (square foot) by 20 inches, which equals 7.2. Area of 228 × 7.2 = 1641.6 brick (rounds to 1,642).

Notes

Notes

Calculate Square Yards Answers/Solutions

1. (a) 20 × 45 = 900 square feet. Divide the area (square feet) of 900 by 9, which equals 100 square yards.
2. (b) If the roll of carpet comes in 12-foot rolls, calculate 30' × 12, which equals 360 square feet. Divide square feet by 9, which equals 40 square yards. Double this to take care of the 20-foot width.

Calculate Cubic Yards Answers/Solutions:

1. (b) L × W × D ÷ 27. The L × W or square feet, has been provided. Multiply the square feet of 875 by 0.5 (6" ÷ 12 = .5) and divide by 27. The answer is 16.20.
2. (c) L × W × D ÷ 27. Begin by determining the area (L × W) of the tank's base. The formula for computing the area of a circle is pi R squared. The radius is half of the diameter or 9'. Pi (3.14) multiplied by the radius squared (81) is 254.34. Multiply the area of 254.34 × 32' for a total of 8,138.88. Divide 8,138.88 by 27 for a total of 301.44.
3. (c) L × W × D ÷ 27. Plug the proper numbers into the formula. The length is 200, the width is 1.17 (14" ÷ 12), and the depth is 1.33 (16" ÷ 12). Multiply 200 × 1.17 × 1.33, and divide by 27. The total cubic yards is 11.52. Round to the next whole number which is 12. Multiply 12 by $65, and the answer is $780.00.
4. (b) L × W × D ÷ 27. Plug in the numbers. Length equals 29, and width equals 14. To determine the depth, subtract 96 from 98; the distance between grade and the top of the slab is 2 feet. Subtract the thickness of the slab from 2 feet. 2' – 0.333 = 1.67. The depth of the sand to be placed is 1.67. Plug this into the formula, and multiply 29 × 14 × 1.67, which equals 678.02. Now, divide this total by 27. The answer is 25.11.

Calculate Cubic Yards Answers/Solutions

1. (d) Begin by sketching an additional 2 feet around the perimeter of the building. Divide the structure into two sections. The first section will measure 54' × 38'. The second section will be 24' × 6'. Add the square feet of both areas, for a total of 2,196 square feet. This is the L × W (area) portion of the formula. Next, multiply this by the depth of 3 feet and divide by 27. The cubic yards to be excavated is 244. Take 15% of 244, and add it to the total. The correct answer is 280.6.

2. (b) L × W × D ÷ 27. Begin by multiplying the area (L × W) by the depth of 7 feet: 4,000 × 7 = 28,000. Divide 28,000 by 27 for a total of 1,037.037 cubic yards. Increase this by 20% to account for the swell, which equals 1,244.444. Divide the total of 1,244.444 by 20 (what a truck holds), and the answer is 63 trucks.
3. (c) L × W × D ÷ 27. Plug the proper numbers into the formula. The length is 72, the width is 14, and the depth is 4 feet. Multiply 72 by 14 by 4, and divide by 27. The total cubic yards is 149.33. Divided by 0.90 (100% − 10% = 90%), the answer is 165.925.
4. (a) Begin with adding 5 feet around the perimeter of the building. The new dimensions will be 62' × 43'. Plug the numbers into the formula, L × W × D ÷ 27. 62' × 43' × 2' ÷ 27 = 197.48. Now, divide by 0.85. The answer is 232.33.

Wall Framing Math Answers/Solutions:

1. (c) Quantity = Wall length ÷ on-center spacing + 1. Multiply 25 by 12 to convert the length of the wall to inches. 25 × 12 = 300. Divide the length of the wall by 16. The studs are located every 16 inches. 300 ÷ 16 = 18.75. There are 19 spaces between each stud. Add one stud to the 19 to account for the beginning stud. This wall will require 20 studs.
2. (b) Board feet = Quantity × Name size × L ÷ 12. The quantity is 18, the name size is 2 × 10, the length is 16 feet. 18 × 2 × 10 × 16 = 5760; now divide by 12. 5760 ÷ 12 = 480.
3. (c) Sketch the wall. Draw a dimension line indicating the center of a 3-foot door. The rough opening will be 38½ inches wide. Divide 38.5 by 2 to determine where the jack (trimmer) stud will be placed. 38.5 ÷ 2 = 19.25 inches. Convert the 9 feet to inches. 9 × 12 = 108 inches. Subtract 19.25 from 108 to identify the measurement of where the rough opening will begin. 108 − 19.25 = 88.75 inches.
4. (c) Quantity = LF ÷ on-center spacing + 1. First, convert the 24 feet to inches: 24 × 12 = 288 inches. Next, divide 288 inches by 16 inches. 288 divided by 16 = 18. Add 1 for the first joist: 18 + 1 = 19 joists.

Roofing Math Answers/Solutions:

1. (a) Divide 28 by 2 to determine the middle point of the span. Multiply 14 by 6 because the roof rises 6 inches for each foot: 14 × 6 = 84 inches. The choices provided are in feet; therefore, we must divide 84 by 12 to convert to feet: 84 ÷ 12 = 7. The height of the gable is 7 feet.

Notes

Notes

2. (c) Pitch is a ratio of rise over span. The 4 in ¼ represents the span. We need to determine the slope. A slope is a ratio of rise to run. If 4 is the span, half of 4 will represent the run. ½ is the slope of this building, but we need it in a ratio to units of 12. Place an *x* over 12, and beside it place 1 over 2. Cross multiply: $12 \times 1 = 12$. Divide 12 by 2. The ratio of rise to run is 6. The answer is 6:12.
3. (d) Run is half of the span: $29 \div 2 = 14.5$.
4. (b) Begin by determining the gable height: 32 ÷ two will indicate the center point (ridge) to be 16. Multiply 16 by 8 (the roof is rising 8 inches for each foot) to determine that gable height is 128 inches. Convert the 128 inches to feet: $128 \div 12 = 10.67$. Multiply 16 by 10.67 to determine the area: $10.67 \times 16 = 170.72$. This accounts for the total area of the gable.
5. (b) The rafters will run to the ridge. Half of the 20 foot width (to the ridge) is 10 feet. Add 1 foot for the overhang: 11×1.054 (roof multiplier for a 4:12 slope) $= 11.594$. Round to the next even number: the answer is 12.
6. (c) Add 1 foot of overhang for each side of the roof: this equals 32 feet. Multiply 32 by the roof multiplier for a 6 in 12 slope roof. This multiplier is 1.118: $32 \times 1.118 = 35.776$.
7. (b) C = square root of A squared plus B squared. $A = 13.5$ (half of 27), $B = 6$: $13.5 \times 13.5 = 182.25$, $6 \times 6 = 36$. Add these two numbers: $182.25 + 36 = 218.25$. Press the square root symbol, and 14.77 is returned.

8. (d) Begin by sketching the overhang around the perimeter of the building. Section A's new dimensions will be 77' × 32' Calculate the area of this section: 77 × 32 = 2,464 square feet. To determine the dimension of section B, begin with the 75-foot length of the overall measurement for the building and subtract 39 feet and 20 feet. Section B is 16 feet in width. Section B's new dimensions, with overhang, will be 18 feet × 20 feet. Calculate the area of this section: 18 × 20 = 360. Add the two sections together: 2,464 + 360 = 2,824 square feet. Increase this by the multiplier, for a 4 in 12 slope. The multiplier is 1.054: 2,824 × 1.054 = 2,976.496. Add 10% for waste: 2,976.496 + 297.6496 = 3,274.1456. The closest answer is 3,274 square feet.
9. (d) Begin by dividing 16 by 2 to identify the run for this section. Add the overhang: 8 + 1 = 9. Multiply by the appropriate multiplier. The multiplier for a 6 in 12 slope is 1.118: 9 × 1.118 = 10.062. Although 10 feet might work, you are required to round to the next even number. The answer is 12.
10. (d) Begin by sketching the overhang around the perimeter of the building. Section A's new dimensions will be 77 feet × 32 feet Calculate the area of this section: 77 × 32 = 2,464 square feet. Section B is 16 feet in width. Section B's new dimensions, with overhang, will be 18 feet × 20 feet. Calculate the area of this section: 18 × 20 = 360. Add the two sections together: 2,464 + 360 = 2,824 square feet. Increase this by the multiplier, for a 5 in 12 slope. The multiplier is 1.083: 2,824 × 1.083 = 3,058.392. Add 5% for waste: 3,058.392 + 152.9196 = 3,211.3116. Divide this total by 100 to convert the square feet to squares: 3,211.3116 ÷ 100 = 32.1131. If there are three bundles needed per square, multiply the squares by 3: 32.1131 × 3 = 96.339. 97 bundles of shingles.

Notes

Notes

Part 2: Administration Answer Key

	IRC	IBC
1. d	106.2	106.2
2. d	105.5	105.5
3. a	106.3.1	106.3.1
4. b	105.7	105.7
5. c	105.2	105.2
6. d	111.2	111.2
7. d	110.3	110.2
8. c	114.1	114.2
9. b	106.5	106.5
10. c	106.5	106.5

Part 3: Design and Planning Answer Key

	IRC	IBC
1. a	310.1.3	1025.2.1 1026.2.1 (2006)
2. c	309.1	406.1.4
3. c	Table 301.5	Table 1607.1
4. b	312.2 Exc #2	1012.3 1013.3 (2006)
5. d	311.6.1 See Exc (2006)	1010.2
6. d	319	2304.11.2.1
7 b	311.5.2	1009.2
8. d	315.2	803.1
9. b	304.3	1208.1
10. d	303.1	1205.2 (8% of 100 square feet)
11. c	305.1	1208.2
12. b	314.2.6 314.5.9 (2006)	2604.2
13. a	310.1.1	1025.5.1
14. c	315.1	803.1
15. b	310.2	1025.2 1026.5.1 (2006)
16. c	309.4	406.1.3
17. a	311.3	1020.1 1021.2 (2006)
18. b	312.1	1012.1 1013.1 (2006)
19. b	311.5.4	1009.4
20. b	309.2	406.1.4
21. d	308.4 #11	2406.3 #11
22. c	311.5.6.1	1009.11.1 1012.2 (2006)
23. b	314.2.3 314.5.4 #6 (2006)	2603.4.1.6
24. b	319.1	2304.11.2.5 2304.11.2.6 (2006)
25. c	317.2.2	704.11.1

Notes

Notes

Part 4: Foundations Answer Key

	IRC	IBC
1. c	Table 404.1.1(1)	Table 1805.5 (1)
2. c	Table 401.4.1	Table 1804.2
3. b	404.1.6	N/A
4. d	Table 402.2	Table 1904.2.2(2)
5. d	403.1.4	1805.2
6. a	403.1.6	2308.6
7. a	401.3 exception	1803.3
8. a	408.2	1203.3.1
9. b	405.1	1807.4.2
10. c	Table 402.2	1805.4.2.1
11. b	403.1.5	1805.1
12. a	403.1.4.1 exception	1805.2.1 exception
13. b	404.1.5	1805.5.1.1
14. b	404.1.5.1	1805.5.7
15. d	404.1.8	1805.5.1.3

Part 5: Floors Answer Key

	IRC	IBC
1. b	502.6.1	2308.8.2
2. b	502.10	2308.8.3
3. d	502.3.1(2)	2308.8(2)
4. b	502.8.1	2308.8.2
5. c	502.12	717.3.3
6. c	502.10	2308.8.3
7. b	506.1	1911.1
8. a	Table 503.2.1(1)	Table 2304.7(3)
9. d	Table 502.3.1(1)	Table 2308.8(1)
10. d	502.6	2308.7
11. c	502.6.2	2308.8.2
12. b	502.8.1	2308.8.2
13. a	Table 502.5(2)	Table 2308.9.6
14. c	506.2.3	1911.1 1910.1 (2006)
15. a	503.3.2	2303.1.7.1
16. b	Table 502.3.1(1)	Table 2308.8(1)
17. d	502.8.1	2308.8.2
18. c	Table 502.3.1(1)	Table 2308.8(1)
19. b	Table 502.3.1(1)	Table 2308.8(1)
20. a	502.8.1	2308.8.2
21. b	Table 502.5(2)	Table 2308.9.6
22. b	502.7.1	2308.8.5
23. d	502.11.3	2308.10.7.3 2303.4.1.7 (2006)
24. d	506.2.3	1911.1 1910.1 (2006)
25. c	502.3.1	Table 2308.8(1)

Notes

Notes

Part 6: Walls Answer Key

	IRC	IBC
1. b	602.3.3	2308.9.2.2
2. c	Table 602.3(5)	Table 2308.9.1
3. c	603.3.3	2211.2.2(7) Refers to AISI (2006)
4. b	Table 602.3(4)	Table 2308.9.3(5)
5. c	602.8	717.2.2
6. c	602.9	2308.9.4
7. d	602.6.1	2308.9.8
8. d	602.6	2308.9.11 2308.10.4.2 (2006)
9. b	Table 603.2(2)	2211.2.2 Refers to AISI (2006)
10. c	603.2.4	2211.2.2 Refers to AISI (2006)
11. b	Table 603.2(1)	2211.2.2 Refers to AISI (2006)
12. a	602.11.3 #1	2308.11.3.2
13. c	602.10.1.1	2308.3.1
14. a	602.6 #1	2308.9.11 (40% of 3.5")
15. c	Table 602.3(5)	Table 2308.9.1
16. b	Table 602.3(4)	Table 2308.9.3(5)
17. a	602.7.2	2308.9.7
18. b	602.5	2308.9.2.3
19. a	602.9	2308.9.4
20. c	602.3.2	2211.2.2.1
21. d	603.2.4	2211.2.2.1
22. a	Table 602.3(1)	Table 2304.9.1
23. b	602.8.1	717.2.1
24. b	602.6	2308.9.10
25. b	602.3.1	2308.9.1

Part 7: IRC Wall Coverings Answer Key

1. b R702.2.1
2. d R703.5.3
3. a R703.4 Table
4. b R703.6.2.1
5. c R702.2
6. c R703.8 #3
7. a R702.4.2
8. d R702.3.5 Table
9. d 1926.56 (Table D-3)
10. b R703.7.3

IBC Wall Coverings Answer Key

1. d 1405.6
2. c 1405.2 Table
3. d 1405.11.1
4. b 1405.9.1
5. a 2502 Definition
6. c 1405.4 #3
7. b 1405.13.1
8. a 1405.13.1
9. d 1405.3
10. d 1405.17.2

Notes

Notes

Part 8: Roof Framing Answer Key

	IRC	IBC
1. c	802.8.1	2308.8.5
2. b	Table 802.4(1)	Table 2308.10.2(1)
3. b	802.6	2308.10.4.1
4. b	Table 802.5.1(1)	Table 2308.10.3(1)
5. b	Table 802.5.1(1)	Table 2308.10.3(1)
6. b	802.7.1	2308.10.4.2
7. b	Table 802.5.1(9)	Table 2308.10.4.1
8. c	Table 802.5.1(9)a	Table 2308.10.4.1a
9. b	Table 802.11	Table 2308.10.1
10. b	802.3.2	2308.8.2
11. d	802.5.1	2308.10.5
12. b	802.3	2308.10
13. c	802.9	2308.10.4.3
14. c	Table 803.1	Table 2304.7(1)
15. b	802.7.1	2308.10.4.2
16. c	802.5.1	2308.10.5
17. a	Table 802.4(2)	Table 2308.10.2(2)
18. d	Table 802.4(2)	Table 2308.10.2(2)
19. a	807.1	1209.2
20. d	802.1.3.1	2303.2
21. a	806.2	1203.2
22. b	802.1.3.5	2303.2.5
23. c	802.1.3	2303.2
24. a	806.3	1203.2
25. a	Table 802.4(1)	Table 2308.10.2(1)

Part 9: Roof Coverings Answer Key

	IRC	IBC
1. a	R905.2.2	1507.2.2
2. d	R905.2.5	1507.2.6
3. b	R905.2.5	1507.2.6
4. b	R905.2.6	1507.2.7
5. c	R905.2.7	1507.2.8
6. c	R905.2.7.2	1507.2.8.2 1507.2.8.1 (2006)
7. b	R905.3.3.2	1507.3.3.2
8. d	R905.2.8.1	1507.2.9.1
9. c	R905.2.8.2 exception	1507.2.9.2 exception
10. a	R905.3.2	1507.3.2
11. c	R905.3.6	1507.3.6
12. d	R905.3.7	1507.3.8
13. a	R905.4.3	1507.5.3
14. c	R905.4.6	1507.5.6
15. a	R905.7.1	1507.8.1
16. d	R905.6.5 Table	1507.7.5 Table
17. d	R905.8.6	1507.9.7
18. b	R905.8.6 Table	1507.9.7 Table
19. d	R905.10.4	1507.4.4
20. c	R905.10.3 Table	1507.4.3 Table 1507.4.3(1) Table (2006)
21. b	R905.5.5.2	1507.6.2
22. d	R905.9.1	1507.10.1
23. c	R905.14.3	1507.14.3
24. b	R907.3 exception	1510.3 exception
25. a	R905.9.2 Table	1507.10.2 Table

Notes

Notes

Part 10: Chimneys and Fireplaces Answers Key

	IRC	IBC
1. b	R1001.1.1	2113.2
2. d	R1001.3	2113.6
3. c	R1001.7	2113.10
4. c	R1001.8.1	2113.11.1
5. b	R1001.6.1	2113.9.1
6. b	R1001.6	2113.9
7. a	R1001.9	2113.12
8. d	R1001.11(1) Table	2113.16(1) Table
9. c	R1001.14	2113.18
10. a	R1001.15	2113.19
11. c	R1001.12.2 Table	2113.16 Table
12. d	R1001.16	2113.20
13. b	R1003.2	2111.2
14. d	R1003.4.1	2111.4.1
15. a	R1003.6	2111.6
16. c	R1003.7	2111.7
17. d	R1003.8	2111.8
18. b	R1003.8.1	2111.8.1
19. a	R1003.9.2 exception	2111.9.2 exception
20. d	R1003.10	2111.10
21. b	R1005.2	2111.14.3 2111.13.3 (2006)
22. b	R1001.10	2113.14
23. c	R1001.11(2)	2113.16(2)
24. a	R1001.15	2113.19
25. c	R1001.2	2113.5

Part 11: Concrete Answer Key

Contractor's Guide to Quality Concrete

1. d Page 73
2. a Page 72
3. c Page 98
4. b Page 68
5. c Page 17
6. b Page 14
7. c Page 66
8. d Page 15
9. c Page 48
10. a Page 66, 74
11. c Page 30
12. d Page 15
13. b Page 81
14. a Page 17
15. c Page 18, 20
16. b Page 19
17. b Page 47
18. d Page 36
19. b Page 30
20. a Page 17
21. c Page 57
22. c Page 127
23. a Page 91
24. a Page 98
25. b Page 20

Notes

Notes

Part 12: Masonry Answer Key

Modern Masonry

1. b Page 134
2. b Page 86
3. d Page 133
4. d Page 73
5. c Page 173
6. d Page 133
7. c Page 84
8. a Page 79
9. d Page 104
10. d Page 107
11. b Page 113
12. c Page 181
13. a Page 135
14. c Page 132
15. b Page 133
16. b Page 107
17. a Page 72
18. d Page 80
19. d Page 85
20. a Page 81
21. a Page 72
22. b Page 73
23. c Page 73
24. c Page 138
25. a Page 72

Part 13: Carpentry Answer Key

Carpentry and Building Construction (6th Edition)

1. b	Page 65
2. d	Page 361
3. d	Page 586
4. c	Page 578
5. b	Page 222
6. c	Page 226
7. a	Page 376
8. c	Page 226
9. a	Page 686
10. b	Page 117
11. a	Page
12. d	Page 222
13. c	Page 226
14. c	Page 589 2" for rough opening plus 1½" for each of the supporting trimmers
15. a	Page
16. b	Page 300
17. b	Page 281
18. b	Page 238
19. b	Page 60
20. b	Page 492
21. a	Page 490
22. a	Page 425
23. a	Page 314
24. d	Page 373
25. c	Page 70

Notes

Notes

Part 14: Roofing Answer Key

	Carpentry and Building Construction	Roofing Construction and Estimating
1. c	Page 595	Page 18
2. d	Page 425	Page 14
3. a	Page 619	Page 405
4. b	Page 595	Page 111
5. b	Page 595	Page 10
6. a	Page 598	Page 309
7. d	Page 620	Page 24
8. b	Page 425	Page 25
9. a	Page 595	Page 111
10. c	Page 611	Page 428
11. c	Page 601	Page 61
12. b	Page 596	Page 79
13. a	Page 602	Page 64
14. b	Page 595	Page 111
15. d	Page 603	Page 96
16. a	Page 424	Page 10
17. d	Page 481	Page 15

Explanation:
1. Take half of the span (24 divided by 2 5 12)
2. Add the overhang to the run of 12 (12 1 1 5 13)
3. Multiply 13 by the 1.118 (roof-slope factor) 5 14.534
4. Round to the next "even" number.
5. Answer is 16

18. d	Page 490	Page 96
19. c	Page 611	Page 15
20. c	Page 607	Page 106

21. a	Not answered	Page 385
22. d	Not answered	Page 390
23. b	Not answered	Page 40
24. c	Not answered	Page 304
25. c	Not answered	Page 301

Notes

Notes

Part 15: Project Management Practice Exam 1 Answer Key

1. a
2. c
3. c
4. a
5. c
6. d
7. c
8. a
9. c
10. c
11. d
12. a
13. c
14. d
15. b
16. d
17. d
18. b
19. d
20. d
21. a
22. a
23. a
24. a
25. a

Part 15:
Project Management Practice Exam 2 Answer Key

1. b
2. a
3. b
4. c
5. a
6. b
7. c
8. a
9. b
10. a
11. b
12. a
13. b
14. b
15. c
16. c
17. a
18. b
19. b
20. b
21. b
22. d
23. a
24. a
25. a

Notes

Notes

Part 16: OSHA Practice Exam 1 Answer Key

1. c 1926 Subpart P, Appendix B (Table B-1)
2. b 1926.1053 #13
3. b 1926.302(a)(2)
4. c 1926.651(J)(2)
5. b 1926.102
6. b 1926.24
7. c 1926.651(c)(2)
8. d 1926.651(h)(1)
9. a 1926.56(Table D-3)
10. a 1926.50(d)(2)
11. c 1926. 651(C) (2)
12. c 1926.302(b)(4)
13. c 1926 Subpart P, Appendix A
14. a 1926.55(a)
15. a 1926.501(b)(1)
16. d 1926.300(d)(3)
17. d 1926.25(b)
18. d 1926.1053(b)
19. d 1926.651(c)(2)
20. c 1926 Subpart P, Appendix A
21. b 1926 Subpart P, Appendix A
22. b 1926.1053(b)(1)
23. b 1926 Subpart P, Appendix A
24. b 1926 Subpart P, Appendix A
25. d 1926.28

Part 16: OSHA Practice Exam 2 Answer Key

1. b 1926.52 (Table D-2)
2. c 1926.1052(c)(4)iii
3. c 1926.151(c)(5)
4. d 1926.302(e)(2)
5. c 1926.104(f)
6. d 1926.451(e)(2)ii
7. c 1926.250(b)(8)iv
8. c 1926.252(b)
9. b 1926.604(a)(2)i
10. c 1926.451(e)(6)v
11. d 1926.150(c)(1)vi
12. c 1926.104(d)
13. c 1926.250(b)(6)
14. b 1926.50(d)(2)
15. c 1926.104(d)
16. b 1926.251(d)(4)iv
17. c 1926.300(d)(3)
18. c 1926.252(a)
19. d 1926.105(a)
20. c 1926.452(k)(1)
21. c 1926.502(b)(3)
22. c 1926.105(c)(1)
23. a 1926.451(f)(6)(Table)
24. b 1926.1053(b)(1)
25. c 1926.451(e)(3)ii

Notes

Notes

Part 17: Comprehensive Final Answer Key

1. b	IRC 602.3.1	IBC 2308.9.1
2. a	IRC 403.1.6	IBC 2308.6
3. b		
4. a	IRC 105.7	IBC 105.7
5. a		
6. a		
7. b	R1005.2	2111.14.3
8. d	R905.3.7	1507.3.8
9. b	1926.302(a)(2)	
10. d		
11. b	IRC Table 805.5.1(1)	IBC Table 2308.10.3(1)
12. a	R1003.6	2111.6
13. c	IRC Table 301.5	IBC Table 1607.1
14. c	IRC 309.4	IBC 406.1.3
15. c		
16. c		
17. a	IRC 806.2	IBC 1203.2
18. d	R905.8.6	1507.9.7
19. c		
20. c	1926 Subpart P, Appendix A	
21. c	IRC Table 404.1.1(1)	IBC Table 1805.5(1)
22. b	R1001.6	2113.9
23. d	IRC 105.3.2	IBC 105.3.2
24. b	R905.2.5	1507.2.6
25. d	IRC 315.2	IBC 803.1
26. d	IRC 502.6	IBC 2308.7
27. d	IRC 506.2.3	IBC 1911.1
28. d	IRC 802.5.1	IBC 2308.10.5
29. c		
30. b		
31. d		
32. d	IRC 502.3.1(2)	IBC 2308.8(2)

33. b	IRC Table 602.3(4)	IBC Table 2308.9.3(5)
34. b	IRC Table 602.3(4)	IBC Table 2308.9.3(5)
35. c		
36. c	R1001.8.1	2113.11.1
37. d	IRC 502.8.1	IBC 2308.8.2
38. c	IRC 602.3.2	IBC 2211.9.2.1
39. b	IRC Table 802.4(1)	IBC Table 2308.10.2(1)
40. a	R905.9.2 Table	1507.10.2 Table
41. c	R1001.12.2 Table	2113.16 Table
42. c		
43. b	1926 Subpart P, Appendix A	
44. a	IRC 310.1.3	IBC 1025.2.1
45. c	IRC 311.5.6.1	IBC 1009.11.1
46. b	IRC 405.1	IBC 1807.3
47. a	IRC Table 502.5(2)	IBC Table 2308.9.6
48. c	IRC 603.2.4	IBC 2211.2.2
49. a	IRC Table 802.4(2)	IBC Table 2308.10.2(2)
50. d	R905.2.5	1507.2.6
51. a		
52. b	R1001.1.1	2113.2
53. a	R1001.15	2113.19
54. a		
55. b		
56. c		
57. d	1926.25	
58. d	IRC 319	IBC 2304.11.2.1
59. b	IRC 404.1.6	IBC
60. a	IRC Table 503.2.1(1)	IBC Table 2304.7(3)
61. c	IRC 602.10.1.1	IBC 2308.3.1
62. c	IRC Table 805.5.1(1)-a	IBC Table 2308.10.3(1)-a
63. b	R905.3.3.2	1507.3.3.2
64. a	R1003.9.2 exception	2111.9.2 exception

Notes

Notes

65. d		
66. d	IRC 303.1	IBC 1203.4 (8% of 100 square feet)
67. b	IRC 404.1.5.1	IBC 1805.5.7
68. a	IRC 502.8.1	IBC 2308.2
69. b	IRC 602.5	IBC 2308.9.2.3
70. a	IRC Table 802.4(1)	IBC Table 2308.10.2(1)
71. b	R905.5.5.2	1507.6.2
72. c	R1001.14	2113.18
73. d		
74. b		
75. c		
76. d	IRC 106.2	IBC 106.2
77. b	IRC 312.1	IBC 1012.1
78. b	IRC 309.2	IBC 406.1.4
79. a	IRC 403.1.4.1 exception	IBC 1805.2.1 exception
80. c	IRC 502.12	IBC 717.3.3
81. b	IRC 502.7.1	IBC 2308.8.5
82. c	IRC Table 602.3(5)	IBC Table 2308.9.1
83. b	IRC 802.7.1	IBC 2308.10.4.2
84. a	R905.3.2	1507.3.2
85. b	1926.651	
86. b	R1003.2	2111.2
87. a		
88. c		
89. d	IRC 311.6.1	IBC 1010.2
90. a		
91. a	IRC 310.1.1	IBC 1025.2
92. b	IRC 502.6.1	IBC 2308.8.2
93. d	IRC 602.6	IBC 2308.9.11
94. b	IRC 802.7.1	IBC 2308.10.4.2
95. a		

96.	b	R905.8.8 Table	1507.9.7 Table
97.	c	R1001.2	2113.5
98.	c		
99.	a		
100.	c	R905.14.3	1507.14.3

Notes

ABOUT THE AUTHOR

American Contractors Exam Services

American Contractors Exam Services has successfully prepared candidates for state required licensing exams since 1991. A team of full time instructors comprises field experience, training expertise and formal education to formulate an effective and dynamic approach to concisely present building codes, math concepts, and construction theory in a seminar format. The seminars are conducted throughout the United States utilizing multi-media and interactive technology.

Utilizing effective graphics, pictures and bulleted text, complex material is presented in an easy-to-understand format. This simplified approach has led to thousands of success stories and happy clients. One is sure to benefit from any material written or presented by American Contractors Exam Services.

Many of the illustrations featured in this book are used in seminar presentations throughout the country by American Contractors Education Services.